TOWARDS A NEW SCIENCE OF HEALTH

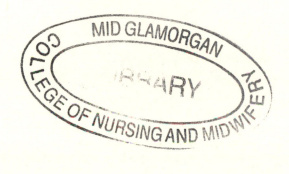

TOWARDS A NEW SCIENCE OF HEALTH

Edited by
Robert Lafaille and Stephen Fulder

London and New York

First published 1993
by Routledge
11 New Fetter Lane, London EC4P 4EE

Simultaneously published in the USA and Canada
by Routledge
29 West 35th Street, New York, NY 10001

© 1993 Robert Lafaille and Stephen Fulder

Typeset in Garamond by
Florencetype Limited, Kewstoke, Avon

Printed and bound in Great Britain by
T.J. Press (Padstow) Ltd, Padstow, Cornwall

British Library Cataloguing in Publication Data

A catalogue record for this book is available from the British Library

ISBN 0–415–08171–8

Library of Congress Cataloguing in Publication Data

Towards a new science of health / [edited by] Robert Lafaille and
Stephen Fulder.
p. cm.
Includes bibliographical references and index.
ISBN 0–415–08171–8 : $59.95
1. Medicine–Philosophy. 2. Medical care. 3. Public health.
4. Medical sciences. I. Lafaille, Robert, 1944– . II. Fulder,
Stephen.
R723.T69 1993
610′.1—dc2093–3282
CIP

CONTENTS

CONTENTS

FIGURES AND TABLES

CONTRIBUTORS

Dirk K. Callebaut is Professor at the University of Antwerp in the fields of plasma- and astrophysics, general relativity and cosmology. He obtained the degrees of licentiate (1957), doctor (1962) and 'Aggregated for Higher Education' ('Habilitation') (1972) in physics at the University of Ghent, Belgium.

David Canter is Professor of Psychology at the University of Surrey. He became Professor of Applied Psychology in 1983 and Head of the Psychology Department in 1987. His original research interest was in the relationship between architecture, buildings and the people that live in them, and he has written seven books on the psychology of architecture and place. He has already researched and written extensively on psychological methods, including the research interview and Facet Theory.

Stephen Fulder received his first degree in biochemistry with chemical pharmacology from Oxford University. He went on to obtain a PhD for work on the ageing process at the National Institute for Medical Research in London. In 1977 he began research into Oriental medicine, alternative medicine and medicinal plants. He established a research consultancy and directed, among other projects, the Threshold Survey of Complementary Medicine, a major research project on the status of alternative medicine in the UK.

Brian Goodwin is Professor of Biology at the Open University. He gained first degrees from both McGill University in Montreal and Oxford University. After research fellowships at McGill (1960–1) and MIT (1961–4) he returned to Britain to take up a lectureship at Edinburgh University and then a Readership in Development Biology at the University of Sussex.

Eberhard Göpel studied medicine and education at the University of Göttingen and wrote a PhD thesis on the social history of health education efforts in Germany. Since 1973 he has been a researcher and lecturer at the University of Bielefeld for health sciences education.

Robert Lafaille co-ordinates the International Network for a Science of

Health. He received his PhD from the University for Humanist Studies in Utrecht for a dissertation on healthy life-style programmes. His main areas of research are health theories, healthy life-style programmes, history of health and life-style, methodology of biographical research, self-help and self-care techniques, and theories of social problems.

Jo Lebeer, MD, worked for a few years as a general practitioner, and then as a counsellor in holistic health promotion with chronic disease patients. From 1988 he has been working as a research fellow at the Centre of General Practice Medicine at the University of Antwerp, where he has carried out research on biographic methodology.

Peter Mielants has an appointment at the University of Antwerp as supervisor in systemic psychotherapy and teacher in communication and psychotherapeutic skills. He treats couples and families in a private practice as a psychiatrist-psychotherapist and individual psychological problems as consultant senior resident in the General Hospital, Stuivenberg.

Lorraine Nanke is a clinical psychologist who developed an interest in complementary medicine while working in the National Health Service. She is currently carrying out doctoral research in holistic treatment evaluation and lecturing at the British School of Osteopathy.

Jens-Uwe Niehoff studied medicine at Humboldt University in Berlin. In 1976 he wrote a doctoral thesis about methodological problems of observational studies in Medical Sociology. His second doctoral degree was awarded in 1984 for a thesis on epidemiological transition. Since 1989 he has been a Professor at the Institute of Social Medicine and Epidemiology at Humboldt University where he became Director in 1991. His research interests focus on social epidemiology, health policy analysis, health promotion and social change.

John Ratcliffe, is an independent consultant in international health, population policy and research methodology for the policy sciences, as well as Adjunct Professor with the Department of Peace and Conflict Studies at the University of California, Berkeley.

Lucas Reijnders has been Professor of Environmental Science at the University of Amsterdam since 1988. His main academic interests currently are waste prevention and energy efficiency. He serves on a number of advisory councils of the Dutch government.

Rudy P. C. Rijke was a co-founder of the Institute of Ecological Health Care in Rotterdam, which provides training courses for health-care professionals to develop awareness of biographical processes, and works on projects with health-care organizations to develop a health and health promotion oriented approach in daily practice.

Paul Rijnders has published on paradoxical communication, medical psychology, psychotherapy and family systems therapy. Since 1991 he has been director of the Zealand Ambulant Psychiatric Institute in Zealand, The Netherlands.

Jack Warren Salmon is Professor and Head of the Department of Pharmacy Administration in the College of Pharmacy, and Professor of Health Resources Management in the School of Public Health at the University of Illinois at Chicago (UIC).

Frank Schneider studied medicine at Humboldt univeristy in Berlin and also at the First Medical Institute in Moscow where the emphasis was on aspects of hygienics. Since 1989 he has been working at the Institute of Social Medicine and Epidemiology in Berlin. His research interests embrace the areas of social epidemiology, prevention policies, health policy analysis and environmental epidemiology.

Janin Vansteenkiste is a sociologist, past lecturer and researcher in the field of sociology of medicine at the Sociology Department of the Catholic University of Louvain (Belgium), at the University of Orange Free State and the University of South Africa. Dr Vansteenkiste is now working as a consultant in management and biography programmes at the Avalon Institute (Ghent, Belgium) where she gives managment courses and does biography work both with groups and individuals.

Marco J. de Vries was Professor at the Erasmus University, Rotterdam and was trained as a Gestalt therapist at the Dutch Gestalt Institute and as a psychosynthesis therapist at the Psychosynthesis Institutes in San Francisco and London. He was involved in training therapists at the London Institute of Psychosynthesis. In 1988 he founded the Helen Dowling Institute for biopsychosocial medicine, of which he is the general and scientific director.

FOREWORD

The WHO's strategy for health promotion has been developing since the early 1980s and represents a key component in the policy framework provided by Health for All.

Thus, in the last decade, the WHO's strategy for health promotion focused on three elements: building and legitimating a framework for health promotion, exploring the dynamic of policy formulation and implementation, and testing such policy in practice at both national and local levels.

The present book is particularly important as regards the first element, 'building and legitimating a framework for health promotion'. In the work of WHO, this element was strengthened by the outcome of three major international conferences. The first held in Ottawa in 1986 produced the Ottawa Charter for Health Promotion. The other two (Adelaide, Australia, 1988; Sundsvall, Sweden, 1991) built upon the process intitiated in Ottawa and contributed to make the concept and principles of health promotion visible and accepted.

In order to promote people's health, research is needed so as to understand, and act upon, the interactions and relationships of individuals with the social, cultural, economic and physical environments within which they conduct their everyday life.

Health promotion research must therefore be dynamic, deal with health as a process and use a multidisciplinary approach. One of the main principles for health promotion research is to focus upon the conditions for change. The identification of factors that facilitate or hinder progress towards environments and individual and group behaviours more conducive to health should be seen as a key goal for research in health promotion.

Furthermore, research is needed to examine the scope and impact of social networks in different contexts. More conceptual and empirical work is required to accurately investigate how social networks influence people's health perceptions and life-styles. This work is needed in order to better understand how different individuals and social groups assess their problems, priorities and resources for actions conductive to health. Some of these issues have been taken up in a recent publication of the WHO Regional Office, Health Promotion Research, *Towards a New Social Epidemiology*.

In health promotion, and more broadly in the rethinking towards the new public health, we need to raise questions that challenge research to move from traditional disease-centred models to health as a resource of everyday living. In this process, it is my hope that the work of the Foundations of the Health Sciences will be instrumental in the years to come in both raising crucial questions and testing new concepts and theory.

Ilona Kickbusch, PhD
Director, Lifestyles and Health,
World Health Organization, Regional Office for Europe

INTRODUCTION

During the last decade a radical shift in views about health and health care has emerged in Western societies. These new views have been given different labels such as holistic health, health promotion, health consciousness, the salutogenetic approach, ecological health, etc. However they all express a deep need to reformulate the subject matter of the health sciences. Proponents view with alarm the strictly curative, technical, impersonal, mechanistic and expensive nature of modern medicine. They view health and disease much more in terms of relationships – a person to themself and to their physical and social environment. The underlying assumptions of health care should be a person's essential human-ness, individuality and indivisibility. This implies a fundamental reform of the health-care system.

These new views emerged as the result of several convergent social trends. First of all serious new health issues and problems arose. Western populations are growing older, and this is accompanied by an increase in the amount of degenerative diseases such as cancer and heart disease; unforeseen side-effects of the health-care system itself (iatrogenesis) emerged, such as thalidomide; new ethical and other problems were caused by advanced medical technology like genetic engineering or advanced medical surgery; increased length of life brought problems of quality of life; ecological health problems arose; modern health care became a heavy and even impossible burden, especially in the Third World. Many of these problems are global in nature and linked to the social, economical and cultural organization of societies and the international community. They cannot be explained and surely not solved by traditional ways of thinking. They also gave rise to new social, moral and political questions (Kickbusch 1989).

Secondly, new grass-roots social movements appeared which challenged the medical and health concepts which have dominated and supported Western health care for so long: the women's movement, alternative psychiatric movements, complementary medicine, self-help groups, consumer awareness, etc.

Thirdly, there were intellectual developments which criticized the dominant scientific theories and notions about the world and sought to replace them. In particular, new developments in physics and chemistry such as chaos theory, quantum mechanics or self-organizing systems disturbed deterministic concepts. New technologies (computers, holography, scanner technology, etc.) supported these developments and hastened the transition. In the human sciences the interdependence of social, psychological and biological elements led to completely new theories. Medicine and the other health sciences were deeply influenced by these developments in science and began a period of rapid transition.

And last but not least, as a result of deep societal and cultural changes, there has been a crisis of values. We believe that much of the current turmoil in the world expresses the birth pains of a new worldview, which will be more adapted to the new social and political realities we live in, and therefore will be international in character. The health issue is a part of, or even a mobilizing force, in that movement. Western society has been perennially egocentric, unable to confront its many problems (internalization of economy, Third World problems, the environment, redistribution of wealth, bureaucratic leadership, the manipulation of democracy by particular interest groups, etc.). This implies that only a deep crisis can induce a real transformation in the way we think. The new worldview will bear little resemblance to the old, for it will need to provide new basic assumptions for the international community. Such a worldview can only be formulated on a meta-level and will have to encompass the existing national and cultural differences in worldviews. The notion of human rights is a good example of such a cross-national notion which will certainly be a constituent part of the growing global vision. The globe, as seen by astronauts turning around our planet, could be its symbol (Kelley 1988).

These various influences have created a favourable climate for original and unorthodox views on health to emerge. However their creators were not, in general, academics or health professionals. The search for a redefinition of health has been, in general, a living and practical experiment within subcultures (see Illich 1979, Capra 1982, 1986 and Hastings 1981). The academics have been very slow to catch up. It can be said that a serious dialogue on the nature of health has hardly begun within the academic world. Take cardiac prevention as an example. While those directly involved in promoting well-being are advising a radical rethinking of life-style (conventional eating habits and stress reduction), the academic community, after a billion dollars spent on clinical studies, is still wrangling over whether or not it is wise to lower raised cholesterol levels.

The opportunities created by the recent popular accumulation of knowledge on health have so far been missed. As the dogma that health is the absence of diagnosable symptoms is weakening, there seems little attempt or inclination to build improved models. Amid the widespread critique of the century-old tenets of scientific medicine, academic researchers can only stand in shock. Revised concepts still seem very far away. If one polled health scientists as to their views on the nature of health, the results are likely to be vague, ill-considered and uncertain. Most would repeat the well-worn textbook definition of the absence of disease, or the more recent WHO description. The acclaimed definition of health declared by WHO at its famous Alma Ata meeting is: 'a state of complete physical, mental and social well-being, and not merely the absence of disease or infirmity'. This definition has found its way into every school book. Yet it is too utopian for current medical professionals to even consider as a basis for their medical practice. Doctors do not invite patients to consult them on questions of well-being. Nor can one imagine a public health policy founded on the principle of well-being, under current political conditions. It is not only irrelevant, it is already out of date. For example it fails to address the importance of the social and physical environment.

Conscious of these limitations, the WHO European Office supported a small gathering, a kind of informal think-tank, to examine if it were possible to begin to reconstruct an understanding of health in an academic context. Does health actually have a description? Can such a description be developed within academic disciplines and institutions and if so, how? What effect would such a rethinking on health have on research in the health sciences and what would those sciences actually look like after such a metaphysical housecleaning?

At its first symposium at Louvain-la-Neuve in 1989, the International Network for a Science of Health responded to the challenges mentioned above by introducing the idea of a science of health. That does not imply a new academic discipline,[1] but a transformation and reorientation of all existing health sciences towards a more integrated and multidisciplinary approach to the problems they are studying. So, when the notion of a science of health is used in this book, it refers first of all to *developments in this direction*. Secondly it indicates a systematic approach to the history, definition and theories of health. Although the development of a science of health faces a multitude of obstacles, we are convinced that the highest priority should be given to a conceptual and theoretical reformulation of our views of health. It has some profound practical consequences in fields such as general practice and public health. There is an urgent need for a fundamental revision of the old health themes (Kickbusch 1989).

THE SEARCH FOR THE REDEFINITION OF HEALTH

In this book various thinkers and researchers within the health field, most of whom participated in the Network's symposium on the science of health (Foundations . . . , 1989), search for such health themes. They ask if the required reformulations of health are possible within the disciplined framework of current research. The answer from all the authors is that a reformulation of our concepts of health is possible, but it is only achievable by a major revision of metaphysical assumptions and modes of enquiry. The authors bring a richly diverse range of backgrounds to this problem. Here we will briefly introduce each contribution and summarize its general approach. Then we will attempt to draw out common themes as a focus of attention.

In the foreword, Dr Ilona Kickbusch explains how a new science of health is required so as to create a more dynamic concept of health, able to relate health as a process, and therefore underpin WHO's efforts at health promotion. WHO is involved in the social dimension of health. It requires more open research models, able to incorporate the social forces that give rise to health and disease.

Part I is general and introductory. It focuses on the main theoretical questions for the development of a science of health. In the first contribution, Dr Robert Lafaille describes some basic frames of reference in the health sciences of today and evaluates them in a critical way. A fourfold classification of theories is presented: the biomedical, the existential–anthropological, the systems and the culturological paradigm. Each paradigm generates its own health definitions and its own type of advice or recommendations on how to lead a healthy life. Divergent concepts of health may lead to contradictory health prescriptions. This poses fundamental questions for the health sciences, and the search for a way to resolve these contradictions is a major challenge to the new science of health. One important way to solve this problem is to construct 'meta-models' which offer a possibility to integrate existing health theories on a higher analytical level. The reader can use this contribution to orient him/herself within the vast area of concepts, definitions and theories in the health sciences. It also gives some background information as an introduction to other contributions.

In Part II a historical analysis helps the reader put all the various views of health in perspective. Dr Eberhard Göpel describes how health views are related to the development of civilization and to broader cultural and social forces.

In Part III ways in which different academic disciplines can be part of a new science of health are explored. Professor Brian Goodwin reveals that a new understanding of living organisms is growing within biology. Organisms can be visualized as developing systems based upon the *field concept*. This frame of reference tries to avoid the cartesian mind–body split in biology. The new biology redefines itself as a science of qualities. Biological research using the field concept allows a new perspective for the development of biology as a health science.

Since the seventeenth century, the concepts of physics have had a deep impact upon science. In addition, they unconsciously filter in to our worldview. Professor Dirk Callebaut explores how physics may contribute to a science of health. Axiomatization, the development of new technologies, new ideas on biological energy, cosmological questions, and questions related to basic dimensions of our world such as time, are assessed for their impact on our notions of health.

Dr Rudy Rijke criticizes the underlying assumption in medicine that health is destroyed or influenced by specific external agents or factors and restored when these factors are neutralized. Yet there is a great deal of evidence that health may be unrelated to pathogenic factors. If you go beyond pathology and examine what is right with people rather than what is wrong with them, a completely different picture of health emerges. Dr Rijke stresses the importance of human autonomy or independence. A science of health should take human autonomy and the dynamic complexity of daily life into account. This need not be in conflict with disease-oriented scientific research, as we know it, but can be complementary, adding a more meaningful dimension.

Systems theory can help to transcend a purely individualistic concept of health and health programmes. The field of human interaction provides a rich source of ideas to a science of health, concludes Dr Peter Mielants in his contribution on theories of human interaction. Systems theory offers a model in which health is an outcome of dynamic interactions between people. In our culture, there is little awareness of the interdependence of phenomena at different system levels.

The development of a science of health will require a dialogue with complementary medicine. Dr Stephen Fulder depicts seven common characteristics or fundamental principles of complementary therapies. These common traits constitute a meta-model which could not only be used to typologize complementary therapies, but also to compare healing practices, across medical systems and even cultures. Complementary medicine enriches and deepens our understanding of health at the individual level. It focuses on questions of vulnerability and disease resistance. It offers knowledge on the elusive grey area between health and disease, the as yet symptom-free 'third state'. It provides data and insight into the constitutional picture of human beings. And it introduces some ingenious methods to measure health. These insights can be used as a background to explore the nature of health and build theories and models to study health processes. Complementary medicine has stimulated new ways of research in the health sciences. New methodologies, such as co-operative enquiry, have been developed to investigate the open-ended, subjective, individualistic and multi-level nature of complementary medicine. They have an obvious value in the development of a new science of health.

The risk-factor approach is dominant in epidemiology. It is highly dependent upon an implicit theory of health which can be summarized by two axioms. The first is that illnesses are caused by determinant factors; the second is that illnesses could be prevented by the removal of these causal factors. According to Professor Jens-Uwe Niehoff and Dr Frank Schneider these two axioms cannot sustain criticism. The health sciences need other, more appropriate theories of prevention. One such theory stresses the fact that all illnesses are unequally distributed in the population and that this inequality reflects the social structure of society. Health becomes a political and moral problem. Professor Niehoff and Dr Schneider attempt to draw a first outline of such a new theory of prevention.

Healing theories are a neglected domain in the health sciences, argues Professor Marco de Vries. What is healing, and how does it proceed? These ought to be key questions for medical and therapeutic practice as well as for the health sciences. Healing is a reaction to the damage and/or distortion of the integrity of the whole system. The healing process repairs this integrity. This can happen for example, by self-regulation, by compensation or transformation of functions or by repair on a higher system level without a restoration of distortions at the lower system level. In some cases new abilities grow to compensate for the damage. In this sense, healing processes might have an evolutionary meaning. This is not only true for individuals, but also for groups, societies or ecological systems. Healing is a function of the total system. This implies that healing processes in the body are mirrored by analogous processes in the mind. It also becomes clear that very limited conclusions can be drawn from the observation of only one level of organization of human beings. A central factor in healing processes is the will to live further and the search for meaning in life. Health promotion can stimulate healing processes by offering people new meaning, and a sense of purpose.

Ecological problems are still growing through a lack of political action. Professor L. Reijnders describes the major ecological problems of our time. The concept of sustainable development is useful as an integrative frame of reference. Sustainable development refers to a stable equilibrium between society and the environment. His analysis shows that modern, industrialized societies are not necessarily threatened by plans to achieve such an equilibrium and solve ecological problems, but can benefit from the development towards a better ecological equilibrium. Health is an example of such a benefit.

The purpose of health policy is very clear: to prevent disease, to reduce death and disability, and to promote health on a community-wide basis. Although scientific knowledge offers an optimistic perspective, the real situation is extremely limited. Professor J. Warren Salmon argues that political culture, economic constraints, and corporate constraints in the health sector itself, are putting serious limits to the effectiveness of public health policies. There have been recent initiatives to create a new public health movement, especially through WHO. In the USA there has been a very limited acceptance of these ideas in comparison with Europe and elsewhere. Economic recession has a negative impact upon public health policy all over the world. A more participative model for public health policy could be a valid tool to break through old patterns and offer a way to empower people to act for their health. A science of health should incorporate such policy considerations.

Shifting views on health and the health sciences would have to be sustained by new methodologies. The development of new methodologies might even have greater strategic value than the development of new theories in the health sciences. In Part IV, possibilities and developments are discussed. For example we stress the need to incorporate personal experience into methods to assess and enquire into health.

Professor John Ratcliffe relates methodology to our paradigmatic viewpoints. He argues that the way to solve the serious global problems we now face depends on the way we ask the questions. An integrative view is outlined which defines health as a state of well-being of man in society. It stresses the social and political context. Research can benefit from such an integrative holistic view. For example, the major improvements in health in this century have been achieved by social rather than medical advances. Health is viewed as a multi-layered phenomenon which can only be investigated by transdisciplinary research within an integrative paradigm. Problem-oriented research should study human well-being at the medical, historical, social, economic, ideological and national levels.

When the symptom is abandoned as the sole measuring rod for health, and attempts are made to include the experience of health in the assessment of health, the usual quantitative methods of medical research become entirely inadequate. Personal experiences are complex, overlapping and multidimensional, and new quantitative methods are required to map them. Professor David Canter and Lorraine Nanke explore this question by concentrating on one method as an example of modelling and analysing data that can include such complex and open systems. They describe how facet theory can be used to reveal relationships between experiences, for example of symptoms, the time frames under which they occurred, expectations of health and treatment, emotional sense of well-being, etc.

The health sciences need to include human experiences, but how are they to do so? One major way is by studying human biography. Biographical research enables a researcher to understand health and disease from the context of a person's whole life. Dr Jo Lebeer draws some outlines for the development of an adequate biographic methodology. Health problems which arise from a deep level of the personality can be studied by biographical research, delving into levels which are not accessible by the more traditional instruments (questionnaire, scales, etc.). Biographical methodology is able to enquire into inner processes and thus reveal the Self and its relationship to health and illness.

In her contribution, Dr Janin Vansteenkiste invites scientists to a dialogue with people and their life experience. It is generally forgotten that health and disease refer to personal experience. A great deal can be learned from life experiences; a knowledge which until now had no place in the scientific discourse. Life experience points mostly in directions other than the main research programmes in the health sciences. For example inner maturity and consciousness, spirituality, the will to be well, and the human need for support in our culture are issues with which a science of health will have to deal. The re-incorporation of personal experience in science implies a shift in the professional attitude of both scientists and health-care workers.

DIRECTIONS FOR A NEW SCIENCE OF HEALTH

There has been a powerful critique of the practice of care-giving within the biomedical system. It is seen as too inhuman, mechanical, disease-oriented, toxic, professionalized, anatomical, etc. An adequate response to this critique not only needs the goodwill of those giving care, but also requires radical new theoretical concepts and paradigms. These would generate a fresh view of the problems and innovative ideas for practice.

This is the aim of a new science of health. The concept of a science of health proposes that a systematic, and comprehensive approach to health be developed which will give answers to the questions raised beyond the biomedical model. A new science of health would question the nature of 'science' as applied to health care. It would invite health scientists and health-care workers to utilize broader-based analyses. Such a new approach will clearly have to be multi- and transdisciplinary in nature and integrate new developments in different fields and disciplines.

The book proposes some ways in which this integration of knowledge might be accomplished. The health landscape can be mapped by means of meta-models which integrate the varied and conflicting models of health arising from culture, history and science (Lafaille, Göpel, Fulder). For example the concepts behind a culturological notion of health, or one derived from the principles of indigenous medical systems can have profound implication for our notions of health. Another direction is the study of complexity (Canter, Ratcliffe, Mielants and Rijnders) and the integration of human experience into our understanding of health and disease. The entire notion of what it means to be healthy, and what it means to be ill can be radically redefined (Rijke). When concepts of health begin to broaden, it soon becomes clear that we have been 'sleepwalking' from within the exclusiveness of the biomedical model. One example is that, strange as it may seem, modern medicine knows almost nothing about the *healing process*. This would be one of the first research projects of a science of health (de Vries). From a theoretical point of view, these would all be important steps towards the development of a non-dualistic view in the health sciences.

Such new concepts have to be complemented by a renaissance in the field of methodology. Methodology has to offer ways to integrate different kinds of knowledge (Ratcliffe). The dualism between qualitative and quantitative research has to be transcended by new techniques which do not destroy the fundamentally qualitative nature of reality (Canter and Nanke). The basic sciences can offer new directions, for example in the search for basic axioms, or in a deeper understanding of processes, fields, energies and uncertainty (Callebaut, Goodwin). Perhaps more important than all of this is the need to revalue human experience and the subjective world in all considerations of health. To allow human experience, personal development, life processes, and the vital importance of healing in the psychological domain to the well-being of individuals, families and societies, a sound biographical methodology has to be created. This methodology, greatly undervalued in modern medicine, is particularly emphasized in this book (Lebeer, Vansteenkiste). It can also be a counterweight to the over-technical nature of the health sciences. All such research will have to acknowledge the influence of the observer on his object of research and make implicit the underlying assumptions of his research.

A new science of health has to be more connected to the health needs of our times. It should be a science of qualities, of human values. This means that it should *include* subjectivity, instead of *excluding* human life and culture. This new science should relate to personal experience, to people's daily lives and even to the tragedy of human existence. This implies a look at the unavoidable moral dilemmas in the health field, such as what constitutes a risk and who defines a risk for whom, what constitutes responsibility, and where are the struggles for health within a divided society (Niehoff and Schneider). All of this has powerful implications for policy. A new science of health will redefine priorities and activities within public health and environmental impact (Salmon, Reijnders).

All the authors are future-oriented. To a certain extent, such an option is risky, because the future evidently remains unknown. Even for experts, it is impossible to gather information on all the new trends and developments. Therefore we do not attempt to give a complete coverage of new ideas that should be incorporated into the study of health as a scientific endeavour. Rather, we give an indication, a series of hints as to the kinds of new perspectives that might be required in a new science of health. The authors as well as the members of the International Network for a Science of Health invite people to join them in the view that there is a fundamental need to reorient the health sciences, to define the health sciences as sciences of human qualities and to develop them in close connection to the evolution of humanity as a whole. A first attempt is made here to sketch out what will be needed for this purpose. If this book succeeds in continuing to stimulate a reorientation in the health sciences, a major aim will be achieved.

REFERENCES

Capra, F. (1982) *The Turning Point: Science, Society, and the Rising Culture,* New York: Bantam Books.

Capra, F (1986) 'Wholeness and Health', *Holistic Medicine* 1: 145–59.

Foundations of a Science of Health, report of a WHO–International symposium held in Louvain-la-Neuve, 19–24 March 1989, Copenhagen: WHO Regional Office for Europe. Unit document ICP/HSR 643.

Hastings, A. C. (ed.) (1981) *Health for the Whole Person*, Boulder, Colorado: Westview Press.

Illich, I. (1976) *Medical Nemesis: the Expropriation of Health*, New York: Marion Boyars.

Kelley, K. W. (1988) *The Home Planet*, Moscow: Mir Publications and Reading: Addison-Wesley.

Kickbusch, I. (1989) *Good Planets are Hard to Find*, WHO Healthy Cities papers, no. 5, Copenhagen: Fadl.

Part I

SCIENTIFIC PARADIGMS AND HEALTH

1

TOWARDS THE FOUNDATION OF A NEW SCIENCE OF HEALTH
Possibilities, challenges and pitfalls*
Robert Lafaille

1 INTRODUCTION

The health sciences have developed several analytical and empirical models or theories. In this chapter we will present a classification of these various insights by distinguishing four paradigms: a *biomedical*, an *existential–anthropological*, a *systems*, and a *culturological* paradigm.[1] Each paradigm generates its own view on health, including health definitions. This richness of models and theories not only leads to manifold research styles and knowledge, but poses many new questions and problems. One important problem is that these different theoretical viewpoints lead to different and even contradictory health advice. This represents a fundamental challenge for the development of a new science of health. There is a need for more integration and synthesis in the health sciences, and this integration will have to be carried out at a meta-level. The different possibilities, dangers and critical aspects of the construction of meta-models will be discussed. The development of a more integrated knowledge will depend on the basic principle that different, even conflicting interpretations of reality can co-exist and are meaningful. It will also have to honour the open and creative character of processes of science and apply this to the development of the health sciences themselves. We have looked carefully at the special position of experiential knowledge. In our opinion the development of a new science of health without a good relationship to human experience will be impossible.

Further, these models and theories are influenced by processes of social and cultural change and by the evolution of worldviews in general. Moreover, the birth of a movement towards a new science of health, and the birth of a new, global worldview are deeply connected. One has to be very aware of this dialectic. It is our conviction that in the future development of the health sciences, the discussion about worldviews[2] and scientific paradigms cannot be avoided. As a consequence of our discussion, a few pathways for the future will be indicated.

1

2 THE MULTIPLICITY OF REALITY: REALITY AS THE BIG UNKNOWN

All science tries to make a logical, linguistic reconstruction of reality. This reconstruction is always made in terms of an existing scientific language and within a major (disciplinary) frame of reference. For centuries, philosophers and scientists thought only one reconstruction of reality was possible. Science, therefore, was conceived as a journey to discover this single reconstruction. It was an adventurous search for 'truth' or 'the laws of the universe'. However it gradually became clear that there could not be a one-to-one relationship between the scientific reconstruction of reality and reality itself. The relationship between science and reality is much more complicated and of a *dialectical* nature. That is, it consists of complex mutual interdependencies.

This dialectical character can be very briefly summarized as follows:

- A multiplicity of different observations are possible. Not only are these observations 'influenced' by reality, but they also have influence on reality itself. By observing reality this reality can be changed. This is especially true for the human sciences.
- There is no archimedic point, no point of a divine observer: the observer is included in the system of observation. Scientific distance is therefore relative.
- The same observations generate an endless series of interpretations; and reversed, every theoretical interpretation/model generates its own observations.
- Scientific knowledge is not only related to time, place and context, but also to history. This includes the biographical processes of the observer as well as those observed. The historical time-dimension might become more and more important as humans penetrate deeper and deeper into the universe.

The more science became aware of the multiplicity of reality, the more that reality presented itself as a big unknown. Instead of revealing the mystery of truth, contemporary science has restored the mystery of reality (see Capra 1975).

3 A POSSIBLE CLASSIFICATION

In principle there is an endless series of possible ways to classify theoretical insights. For example, every individual author can be conceived as representing a different body of propositions, a different look, upon the subject of health science. The level of integration within this classification criterion will remain very low. One can also classify on higher levels of abstraction. That would mean, in particular, gathering separate theories into theoretical

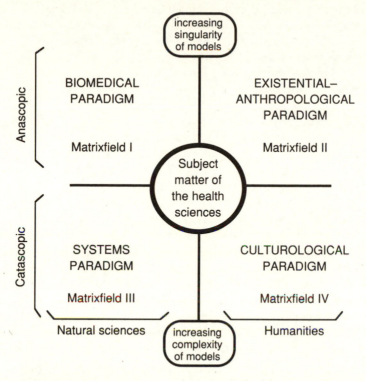

Figure 1.1 Contemporary paradigms in the health sciences

schools. This introduces uncertainties concerning their similarity. The higher the level of classification, the more uncertainty exists concerning the legitimacy of the classification. There is no single solution to this classification problem.

The four-field matrix (Figure 1.1) is proposed to classify the different types of scientific thinking in the field of health science. We distinguish in the field of the health sciences a *biomedical*, an *existential–anthropological*, a *systems*, and a *culturological* paradigm. Theoretically, every paradigm can be formulated in an axiomatic way, using a limited set of basic assumptions which will show a certain coherence. If these paradigms were to be axiomatically reconstructed, they would represent different logical spaces, just like the different forms of geometry (Euclidean, Riemannian, parabolic geometry, etc.). There is a degree of concordance (*Wahlverwantschaft*) between every paradigm and certain types of research, methods of observation, epistemological assumptions, conceptions about causality, or more generally, about the kind of relationships between empirical phenomena, and political and policy issues.

In relation to Figure 1.1, we can make the following statements:

- In the literature a distinction is often made between a so-called 'medical' and 'holistic' model. The medical model only applies to the first matrixfield. The holistic one to the three other matrixfields.
- There is an increase, from top to bottom, of the number of factors which are simultaneously investigated (increase of complexity and synthetic power of theories).
- In the two upper matrixfields, theories are strongly directed to the individual (or his body), in the two lower fields the influence of social relationships and society as a whole receive much more attention. To clarify this distinction, we will use the proposed terms by Zijderveld (1966) *anascopic* and *catascopic*.[3] The term 'anascopic' refers to scientific forms of thinking which strongly focus upon the individual (or its smaller components such as his/her body, an organ, tissues, cells, etc.) and looks from that point of view toward bigger wholes; the term 'catascopic' is used when wider contexts (groups, society, etc.) are taken as a main frame of reference and the individual behaviour is studied and explained in relationship to this broader context.
- The theories in the two fields at the left side are characterized by a positivistic, causal, quantitative and operational style of thinking; the theories at the right side have a more historical, interpretative signature and are much more open to symbols, emotions, meaning in life, etc. The left side is grounded in the natural sciences, the right side in humanities.

Below, we will describe each part of this scheme at greater length, and give an example of research and the kind of health advice developed within the context of each paradigm.

4.1 THE BIOMEDICAL PARADIGM

4.1.1 Definition

The biomedical paradigm has emerged gradually out of the Renaissance and can claim a very glorious genealogy. The central characteristic of this paradigm is the materialization of health and illness. Illness becomes a 'biological fact' and is defined in terms of tissues, organs, cells, nerves, bloodstream, etc. The natural sciences are taken as a model for the medical and the health sciences. The biomedical paradigm is rationalistic and mechanistic (Descartes–Newton). It is based upon dualism, which makes a division between an objective body and a subjectively experienced corporality. Experiential aspects are – usually – declared to be non-specific factors and treated as epi-phenomena. Rationalistic analysis, when carried far enough, can lead towards atomism (cf. the cellular pathology of Virchow) or causal thinking in terms of a singular Stimulus–Response scheme (Newtonian mechanics).

We can further distinguish within this paradigm another two subtypes. The *monocausal* subtype attempts to find one cause or one class of (biomedical) determinants. During the last century monocausal thinking became classical and generally accepted. Nowadays one does not think so much in terms of one cause or one class of determinants (biochemical, physiological, etc.), but in terms of a *multifactorial* causation. This development follows the growth of diverse new academic disciplines (medical psychology, psychotherapy, medical anthropology, medical sociology, etc.) and also the growing possibilities of data-processing (computer). This extension of the paradigm to its multifactorial subtype is certainly an advance, but nevertheless it is still based on all the original postulates of this positivistic, biomedical paradigm, namely its mechanistic, causal and dualistic thinking.

4.1.2 Typical example of research

Phenylketonuria (PKU) is a relatively rare genetic disease which causes a very severe state of mental retardation. It concerns a distortion of the metabolism of phenylalanine, one of the essential amino acids of the human body. Twenty amino acids exist in food, but only a few are essential. Some of the phenylalanine is converted into tryosine, which will be used later on to produce important bodily components such as adrenaline, melanine and thryroxine. This is done by the enzyme phenylalanine-hydroxylase. In PKU this enzyme is deficient as a consequence of a mutation in the corresponding gene. The cause of the illness is thus clearly monocausal. In 1954 Bickel discovered that the severe consequences of this illness, and especially the severe mental retardation, are the result of a toxic accumulation of phenylalanine. It could be prevented when a diet is prescribed from birth in which phenylalanine is almost totally absent (life-style therapy). The concept behind this treatment is obvious: one would prevent the occurrence of the poisoning described above by cutting off the input. Children who receive such a treatment from birth on develop normally, although they certainly show a retardation in physical growth because of the reduction of a necessary amino acid in their food.

4.2   SYSTEMS PARADIGM

4.2.1 Definition

The systems paradigm grew out of simultaneous developments in various scientific disciplines or their fields of application: cybernetics (Wiener), the mathematical theory of information (Cl. E. Shannon and W. Weaver) and the computer (Von Neumann). Generally, Ludwig von Bentalanffy, a biologist of Austrian origin, is referred to as the founder of this paradigm. In

1954 von Bentalanffy founded, together with Boulding, the economist, the mathematician Rapoport and the neurophysiologist Gerard, the Society for General Systems Research, which aimed to develop a general systems theory which would transcend the different scientific disciplines. The ideas of this group were very influential and had a great impact on the work of a large number of scientists.

In this paradigm the concept of a system has a central place. A system can simply be defined as a set of elements between which a specific, non-random, pattern of relationships exists. A system can be separated from its environment; the environment is defined as everything which does not belong to the system but holds a relationship to it. The relationship between system and environment can be described in terms of input and output. It is worth noting that not every input generates an output and that the period of time between input and output can be very long. The system concept is not only applicable to physical phenomena, but encompasses an area of phenomena (inclusive language) as vast as possible.

There is a difference between *open* and *closed* systems. In the case of closed systems the final position of a system (equilibrium) is totally dependent upon the starting position, such as in the case of certain chemical reactions. This is not the case in open systems. In such a system the position of equilibrium is much more dependent upon the characteristics of the whole system instead of being dependent upon specific changes of its internal elements. Examples are the composition of the blood or the temperature of the body. This tendency of open systems to maintain a certain state of equilibrium is called *homoeostasis*.

In general, a system can be divided into subsystems that are mutually dependent and that can be conceived as environments to each other. Sometimes a clear hierarchy between these subsystems can be demonstrated in which functions and processes which are situated at one level use the possibilities and characteristics of other levels. *Reductionism* is an analytical method in which phenomena of a certain level are explained by characteristics of another, lower level or one which is thought of as more fundamental. Thus chemistry was reduced to physics (see the chapter by Callebaut in this book) and, indeed, not without a certain success! And the reduction of physiology to chemistry was for a long period seen as an important scientific goal. Figure 1.2's scheme of the hierarchical system of the human being can further illustrate the basic conceptions of the systems paradigm (according to de Vries 1985: 25–33).

In the systems paradigm the simple Stimulus–Response (S–R) scheme has been elevated to a more complex concept of causality and interdependency between levels, in which cybernetics and its feed-back mechanisms are operating.

Gregory Bateson deserves a special place here. He developed a very open form of systems theory. He is recognized by many contemporary holists as

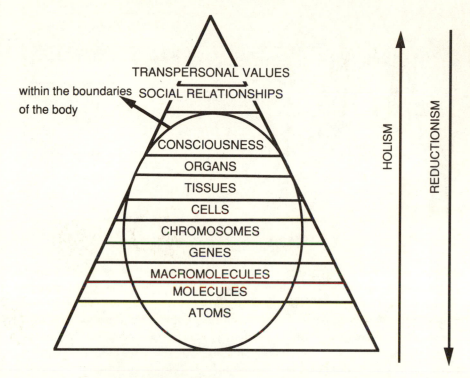

Figure 1.2 Hierarchical levels within the human system

one of their main sources of ideas and he also founded the Palo Alto group which deeply influenced contemporary psychotherapy. Bateson's main contribution is that he offered an escape from rigid causality. In his view, we need a shift in perspective from sciences of objects to sciences of relations. This implies a new scientific language that emphasizes the dynamic character of reality. His main thesis was that there is *a pattern that connects*. He believed that it would be possible to develop non-dualistic ways of thinking. He rejected a split between spirit and matter (Bateson 1972, 1979).

More recently, there has been a shift of attention toward building theories about open systems in which free will and individual choices and less rigid concepts of environment are involved. There have been significant developments in the epistemological foundations (Callebaut and Pinxten 1987), the mathematical and statistical aspects (Thom 1972; Zeeman 1977; Oud 1978; and others) in linking the different system levels to each other (Maturana and Varela 1988) and in situations of disequilibrium (Prigogine and Stengers 1983).

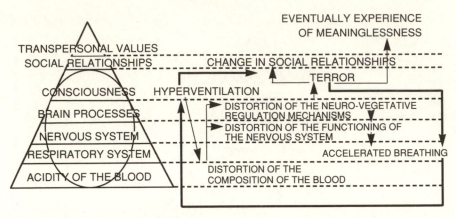

Figure 1.3 Feedback mechanisms in hyperventilation

4.2.2 Typical example of research

Hyperventilation is a change of the respiratory system in which too much carbon dioxide is exhaled (Compernolle 1981). Hyperventilation may occur occasionally (such as when blowing up an air-bed) or chronically. In the case of chronic hyperventilation one speaks of a hyperventilation syndrome, a state which is characterized by frequently recurring fits of terror. These can be very worrying for the patient and also for his environment (partner, family). Carbon dioxide plays an important role in the human body. It influences directly the functioning of the nervous system, regulates the interchange of oxygen in the blood, has a direct influence upon the respiratory regulating centre and helps to hold the acidity of the blood on the right level. The acidity of the blood influences very strongly all the chemical processes in the body, their velocity and their balance. Equilibrium in the chemical household is disturbed by hyperventilation and this might have effects on different levels: headache, heart-fluttering, terror, unclear or double sight, shivers, cramps in arms and legs, etc. We can represent this as shown in Figure 1.3.

Feed-back mechanisms start to operate; the main line is indicated in our scheme by a bold arrow:

• The person suffering from hyperventilation does not as a rule recognize the link between the respiratory changes and his feelings of terror, which, in turn, is a (new) cause of terror.

A first mechanism of reinforcement originates:

increasing terror → accelerated breathing → increasing hyperventilation → increasing symptoms → increasing terror

In the case of chronic hyperventilation this leads to anticipated terror: the fear that there might occur fits of terror.

- The enervation of neurological and endocrinological systems causes bodily symptoms such as acceleration of the pulse, rise in blood pressure, etc. The patient can also experience these symptoms as terrifying. So a second mechanism of reinforcement appears:

  terror → neuro-endocrinical changes → bodily symptoms → terror

  This second mechanism can in turn reinforce the first one.

- A third mechanism appears, through another of the symptoms, namely shortness of breath. If the patient feels that he cannot get enough air, he starts to breathe more heavily, which is the cause of hyperventilation.

By the interaction of these three reinforcing mechanisms a very simple phenomenon can turn into a completely uncontrolled fit of terror with very unpleasant phenomena throughout the whole body. Very often this is accompanied by real fear of death and of becoming crazy. Fortunately the body possesses enough safety mechanisms which are able to cut off these self-maintaining circles long before severe damage can be caused. Therapy can be given on each level:

- the rhythm of the natural breathing can be held or restored by breathing exercises;
- the level of carbon dioxide can be regulated with the use of a plastic bag, or even medication;
- one can learn to cope with fears and terror by psychotherapeutic techniques.

This multiplicity of forms of treatment makes it possible (a) to a great extent to individualize the therapy, (b) to offer the patient an array of choices of therapy, eventually in combination with each other and (c) to avoid the frequently occurring discussion about the 'best' form of treatment: everything that cuts the vicious circles is helpful.

4.3 THE EXISTENTIAL–ANTHROPOLOGICAL PARADIGM

4.3.1 Definition

The existential–anthropological paradigm refers to a systematic body of theoretical concepts which underlie many intellectual schools or traditions. Well known representatives of this paradigm include: psychoanalysts (Freud and adherents), existentialists (like Sartre with *Saint Genet*), symbolic interactionists (Becker, Goffman, Lemert, and others), symbolic medical anthropologists (Devisch, the Canadian School, etc.), modern psychotherapeutic schools (bio-energetics, Gestalt, Psychosynthesis, Neuro-Linguistic-Programming, etc.), phenomenologists (Metz, Coenen, etc.). A certain area of witness literature about health and illness can be regarded as part of this paradigm (Cousins, Geneuglijk, Clabbers, etc.).

In this paradigm there are no firm boundaries between literature, science and personal experience. By 'existential' we mean that the living experience is seen as very important. That is, meaning, structures of significance, consciousness, human values, interpretations of reality and emotions, are recognized as important constituent parts of fundamental knowledge. A qualitative 'humanistic' approach is proposed. This leads in this paradigm to an emphasis on the uniqueness of the individual and his world of living experience. In the field of medicine and health science this implies an appreciation of the individual biography and the significance of life events, crises, emotions, etc. Within this paradigm an abundance of material can be found which concentrate on psychological questions.

This group of thinkers and researchers distance themselves from the positivistic worldview (also called physicalistic objectivism) of the biomedical and systems paradigm. They dispute the division between object and subject that has been advocated by the proponents of this worldview, not only in the human sciences but also in the natural sciences (Capra 1975). A basic assumption is that the human subject is creator of the world in a mutually dependent way with other people.

4.3.2 Typical example of existential–anthropological research

Metz (1975) carried out research in a neurosurgical clinic for patients suffering incurable pain. He came to the following conclusions. The neurosurgeons of the pain clinic make use of a theory of sensory awareness, based on the concepts of René Descartes. Descartes looked at the human being as a robot: when an object penetrates the body, this stimulus is transmitted by the nervous system to the regulating centre, the brain, which in turn responds with a reaction (e.g. by pulling away the limb). The classical, medical models, and especially the physiology of pain, elaborate this basic idea, but deny pain as an experiential reality. Yet it is evident that pain cannot (and should not) be disconnected from the experience of pain. It is absolutely necessary to include this basic quality of pain in each investigation. In the neurosurgical clinic the *experience* of pain is always out of scope. Moreover, medical treatment shifts the attention from the subjective complaint or symptom to a kind of objective fact: the biological deviation (pathology) which is defined as the cause. This transformation has profound consequences for the patient and the therapeutic climate.

The natural science paradigm leads to many contradictions:

- The neurophysiological model cannot aid those cases in which no objective, demonstrable deviation can be found.
- The causal way of thinking of the current neurophysiology cannot

adequately understand the 'changeable' character of the experience of pain, and the strong individual differences in pain experience.

- The causal frame of reference of pain theory implies that the severity of the pain should be proportional to the gravity of the problem. This contradicts empirical observation: there can be severe pain in cases of minor deviations, even without lesion; on the other hand, there are severe lesions which give little or no pain.
- The current pain theory is unable to give a serious explanation for the experience of pain and considers it as an epiphenomenon, a by-product of brain processes. Physiology considers itself unable to investigate such 'shadowy' phenomena. This is not a very consistent attitude for an empirical science!
- The current pain theory gives no adequate explanation for the lack of pain in the case of very serious damage. Very severe damage contracted in the war on the battlefield appeared to be painless if it meant the end of the military career. The meaning which is given by the person to the damage is decisive, and not the biologic severity as such.

An important distinction has to be made between a case history and a life history (Metz 1975; Lafaille and Lebeer 1991). The neurophysiological doctor is only interested in the case history, although pain has to be understood above all as a part of the life history. From the study of the life histories of 250 patients, it appears that the pain experience is a mode of existence, the symptom of a disintegrated life.

From the therapeutic point of view the patient has to be helped to overcome the disintegration of his existence. This disintegration is only in a minor way a property of the person, much more a property of the network of relationships (and their interdependence) in which he participates. Therefore counselling and relation therapy are the indicated pathways. The task of the physician as a consequence changes very fundamentally: in the first place he has to orient his action to the healing of the disintegration instead of restoring the 'deviation'. He also has to concentrate on the healing of the social network to which the patient belongs.

4.4 THE CULTUROLOGICAL PARADIGM

4.4.1 Definition

The culturological paradigm[4] is allied to the existential–anthropological paradigm by emphasizing human experience, but differs from it on core elements. The focus on the individual is broadened to include the societal and cultural roots of health and illness (see Lafaille 1991: Chapter 1). By enlarging the focus it has a certain affinity with the systems paradigm.

The cultural paradigm has its roots in the critique of culture, the sciences of history, macro-sociology and social and cultural anthropology. It is less developed and less systematic than that of the biomedical paradigm or systems theory. Nevertheless many scientists are searching in this direction: psychology of culture (Van den Berg, Campbell, Fortman, Peeters, Wilber), medical history (Verbrugh, Schipperges), certain medical anthropologists (Kleinman, Devisch), sociology of culture (Bourdieu, Beck, Elias, Zijderveld), structuralism (Lévi-Strauss, Foucault), etc. As this paradigm is not so coherent, fruitful insights of other paradigms are incorporated following the personal view of researchers within a broader, social and cultural context.

A main root of the culturological paradigm is the critique of culture. This tries to analyse the suppositions that underlie social life and culture. In relation to health, the dominance of biomedical thinking in the current health system, as well as prevention by promoting healthy life-styles, would be subjected to critical examination. It would expose the hidden (cartesian) dualistic view of man and the reduction of an ill person to an ill body that follows from that view. The social and cultural roots of the evolution toward the dominance of the medical model are analysed in depth. Obviously diseases and complaints, and the way of coping with them, can be related to the existing social and cultural relations in society. Critical objections have been raised towards the spread of preventive measures because they could legitimize processes of social disciplining (Elias 1969; Goudsblom 1977; de Swaan 1978; Niehoff and Wolters 1990). Healthy life-style programmes (such as yoga, macrobiotics, health education programmes, transcendental meditation, etc.) can be seen as an indication of the increasing secularization process in contemporary societies (Cox 1978; Roszak 1976).

The second main root consists of the sciences of history. Historical comparative research is a favourite element of theory building. Comparisons are not only made between populations, but also cross-culturally.

The culturological paradigm stresses, not so much the discovery of universal laws, but that reality is a process. Attempting to fit phenomena into unifying concepts is regarded as a problematic practice which needs critical reflection (see Bourdieu *et al.* 1968). In opposition to the biomedical model, which favours the discovery of universal laws, the culturological paradigm proposes the axiom of the variability of phenomena. What might be a religion today, could be something else tomorrow, despite the terminology which might remain identical. If for the biomedical model variability is a kind of anomaly, dismissed by statistical explanations, for the culturological paradigm the similarity of phenomena is something that needs extra explanation: how, in a pattern of permanent dynamic changes, can invariability occur? The culturological paradigm admits that extensive changes in the view of the world can change the shape of the so-called 'objective reality'. Verbrugh (1983) gives the example of the notion of the body in the Middle Ages which has to be considered totally different from the contemporary

human body, even on the purely biological level. The culturological paradigm aims to cultivate a critical–reflexive attitude within the everyday
practice of research and looks for ways to integrate more explicitly the
social–cultural influence on the researcher in the practice of research.

Investigating the relationship between culture and behaviour has a high
priority in this paradigm. Culture refers to existing and operating values,
conceptions, norms, targets and expectations in society. These have mainly a
dynamic character (patterns) and are collectively shared by people. The
relationship between culture on the one hand and health and illness on the
other hand is complex, dialectical and multidimensional. Culture influences
the definitions of health and illness and the concrete situations in which
these labels are applied, namely the concrete processes of ascription of the ill
role, perceptions of the causal determinants of illness and health, the incidence and prevalence of illnesses, the experience of being ill and the way to
deal with it, coping with emotions and pain, the collective and individual
patterns in relation to care in health and illness, and life-style itself. In the
culturological paradigm the historical aspects of all these factors are strongly
stressed.

4.4.2 Typical example of culturological research

Elias (1969) presents in his two-volume, magistral work, a general theory of
the changes in the behaviour and the emotional life of people of the
Latin–Christian civilization from the Middle Ages up to the present time.
This macro-sociological theory is called by him the civilization theory. The
civilization process is a very broad societal transformation process in which
all parts of society participate. Civilization consists of the gradual development of new behaviour patterns. Elias's basic idea is that there exists a
parallelism between the sociogenesis and the psychogenesis. He means by this
that the development of a personal consciousness, the inner life of the
individual (the emotional housekeeping) and the macro-sociological connections are at a very profound level interdependent. On the basis of an
extensive investigation of historical sources, he shows that there is a shift
from external control (*Fremdzwang*) to self-control (*Selbszwang*). The function of this process is social regulation of the impulses. According to Elias
we can see a vast process of social disciplining, which encompasses hygienic
measures.

Elias proves, with much historical evidence, that in most cases social
motivation for a certain behaviour appears first, later on a hygienic legitimation, and at last enforcement for health reasons. Historically there is
no evidence that hygienic measures in general served medical purposes. On
the contrary they are the result of a process of social disciplining. Most
hygienic behaviour was a mark of social distinction. Further, there is a
strong resemblance between hygienic arguments and moral arguments for

13

promoting certain behaviour. Both have a general 'objective', non-specific meaning.

Scores of authors were inspired in their scientific work by Elias. Here we will only mention two, because they applied the theory of Elias to health and disease. Goudsblom (1977) used the theory of Elias to investigate further the fear of infection, and hygiene. A detailed analysis of the most important diseases, syphilis, leprosy, pestilence and cholera, leads Goudsblom to the conclusion that there has certainly been some influence in the case of syphilis upon the civilization process, but in the case of the other illnesses the reverse is more true, namely an influence of the civilization process upon the spread of these diseases. Nevertheless even in the case of syphilis one has to be cautious with the interpretation of historical material.

De Swaan (1978) applies Elias's theory of civilization to agoraphobia. According to de Swaan a social transformation process occurred which he interprets as an evolution of social relationships from the basis of external command towards relations based upon negotiations. In the nineteenth century strict restrictions of curfew were applied to women. After 1890 these restrictions were loosening and disappeared slowly. Industrialization was accompanied by an enormous transit of population from the country-side towards cities and towns. The aristocracy tried to defend themselves against this condensation of population by personal life guards or by the use of a carriage for transportation; the middle class by increasing restrictions of curfew. The intention was to create a greater physical and social distance (chastity–respectability). This strongly accentuated class differences. Next to these social restrictions appeared different reform movements, which were intended to reduce the annoyance for the lower classes. This societal evolution affected the fantasy and experience of individuals (psychogenesis). The street appeared as a place where people could molest each other or look for sexual contacts: frightening and secretly seducing.

From 1872 on, agoraphobia appeared as a new disease. First this illness appeared within a small group of men, later on women in large numbers became afflicted. The appearance of the new disease is connected with social and economic changes which were the start of a liberation process for women. After 1890 the curfew restriction disappeared very quickly. First, there was a stage of transition: the widening of freedom to go out was given in contexts of segregation: in separate train compartments, separate pubs for women, etc. Women could walk freely on the street, but had to avoid certain times and certain places (prostitution areas). The basis for restriction became physical safety. At that time agoraphobia started as an incomprehensible anxiety about entering public spaces. Psychiatrists of that time interpreted it as a personal phenomenon and defined it as a neurological disorder of the senses, the nervous system or the brain. Psychoanalysts later on interpreted it as an unconscious conflict which stems from childhood. Contemporary psychotherapists look to it as a sign of a disequilibrium between partners.

According to de Swaan agoraphobia can be seen as a result of social taboos of middle-class women in the preceding period. What at that time was strongly recommended (namely not to be alone on the streets) became in the next historical period abnormal and problematic. As a consequence it was treated as a personal inability to move freely through the world.

5 PARADIGMS AND CONCEPTS OF HEALTH AND DISEASE

Every paradigm is connected to certain definitions of health. A comprehensive overview is given in the scheme in Table 1.1.

It is worth mentioning that definitions of health usually cover implicit or explicit utopian ideas. The biomedical concept of health hides the utopian ideal of controlling biological mechanisms by medical knowledge. It is therefore not surprising that the WHO definition, which very clearly is of utopian nature, found so much response. Everybody can find his or her interpretation of health in this definition. With this definition consensus can easily be reached, even at an international level, because this definition refers to utopian and very general principles. Can we give the WHO definition of health a more concrete basis, so that it would point to an empirical reality? Perhaps the most valid way to do this is to develop different concrete utopias which can all be brought together in a democratic process of decision making.

We can now place all the elements together in a scheme (see Figure 1.4). We have seen above that different paradigms are possible, and that they all imply their own definition of health, observational procedures, theoretical models of interpretation (theories), practice and health advice.

One practical problem is that the advice may be radically different or contraditory. For example, disciplinary measures (Matrixfield I) might be in contradiction to the principle of self-realization (Matrixfield II). This diversity will increase in proportion to the development of new theories. There is already a surfeit of health advice which hinders a real and good choice. The same problems were uncovered by research on self-care techniques (Lafaille *et al.* 1981–5; Lafaille 1984). There are so many possible self-care techniques that one needs considerable expertise to cope adequately with this bulk of information. Neither criteria of scientific research, nor leaving this problem to personal choice, are sufficient conditions for a good solution.

6 PATHWAYS FOR THE FUTURE

(a) A first aim for the future is the construction of a meta-model or models of already existing theories. This means to integrate existing knowledge into models of a logically higher order, to solve conceptual dilemmas and to generate new hypotheses. It is a difficult challenge and there are

Table 1.1 Definitions of health

	Biomedical paradigm	Systems paradigm	Existential-anthropological paradigm	Culturological paradigm	WHO
Empirical point of reference	The healthy *body*.	The total system. Every part of the system (subsystem).	Healthy *person*.	Healthy person. Healthy society.	None.
D I Disease phenomena (disease-dimension)	Main focus.	Part of processes within the system (subsystem).	Only important as far as it has a meaning to a person.	Variable: disease is a cultural phenomenon. Varies in different cultures.	Does not apply here.
S E The experience of illness (illness-dimension)	Epiphenomenon.	No special attention. Part of the subsystem mind/consciousness/personality as far as it is related to behaviour.	Main focus.	Experience is the result of the whole cultural context (socialization of social values and belief systems, social control).	Does not apply here.
A S E Sickness behaviour (sickness-dimension)	Only interested in external behaviour in so far as it holds a relationship to disease.	Main focus.	Only important as far as it has a meaning to a person.	Sickness behaviour is the result of cultural processes. Sickness can be used to detect these processes.	Does not apply here.

(cont.)

Table 1.1 cont.

H	*Health phenomena*	Not relevant.	(Dynamic) harmony/equilibrium.	Self-realization. Health as meaning in life. Dynamic.	Health is a main social value and as such a social construction of reality.	Uropian (complete). Static (state).
E	*Health experience*	Not relevant (except as causal factor).	No special attention. Part of the subsystem consciousness/personality as related to behaviour.	Experience is at the core.	Parallelism (dialectics) between collective culture and personal experience.	Well-being. Human experience as criterion for health.
A						
L	*Health behaviour*	Very relevant, but considered only in terms of disease namely the diminishing of risk-factors.	Behaviour that attempts to restore or maintain the equilibrium.	By acting. Corporality.	Collective behavioural patterns are the causes of health and disease. Social and cultural change are necessary to promote health.	Not mentioned.
T						
H	*'Causes' of health*	Absence of disease.	Harmonious functioning of system and subsystem.	Expression of existential problems.	Health and disease reflects the whole culture and the whole society.	Physical, mental and social.
	Healthy life -style programme	Prevention by medical technology.	Biofeedback at regular times (biorhythm).	Personal meaning in life, to solve existential problems and make choices in life; unity of feeling and experience. Every life is unique. Importance of biographical processes.	Change of way of life (individually and collectively). Care for environment and ecology. Change of the total culture is seen as a necessary condition for health for all.	
	The relationship between health and society from the perspective of each paradigm	Out of the normal scope of this paradigm.	Relationship of interdependence. Treated as a technical and policy problem.	Vague, non-specific relationship. Not always mentioned. Dependent upon the belief system and the creativity of the investigator.	Health is an individual *and* a societal problem. Politicization of health problems.	Implicit, but without doubt, a politicization of health and illness.

- to take vitamins
- hygienic control of behaviour
  (brushing teeth, avoidance
  of heat and cold, etc.)
- mastering of health by new
  (medical) technology

- search for meaning in life
- being oneself
- to restore the unity of body,
  emotions and mind
- personal choice
- crises are existential (illness)
  and have to be experienced
  as meaningful events in life

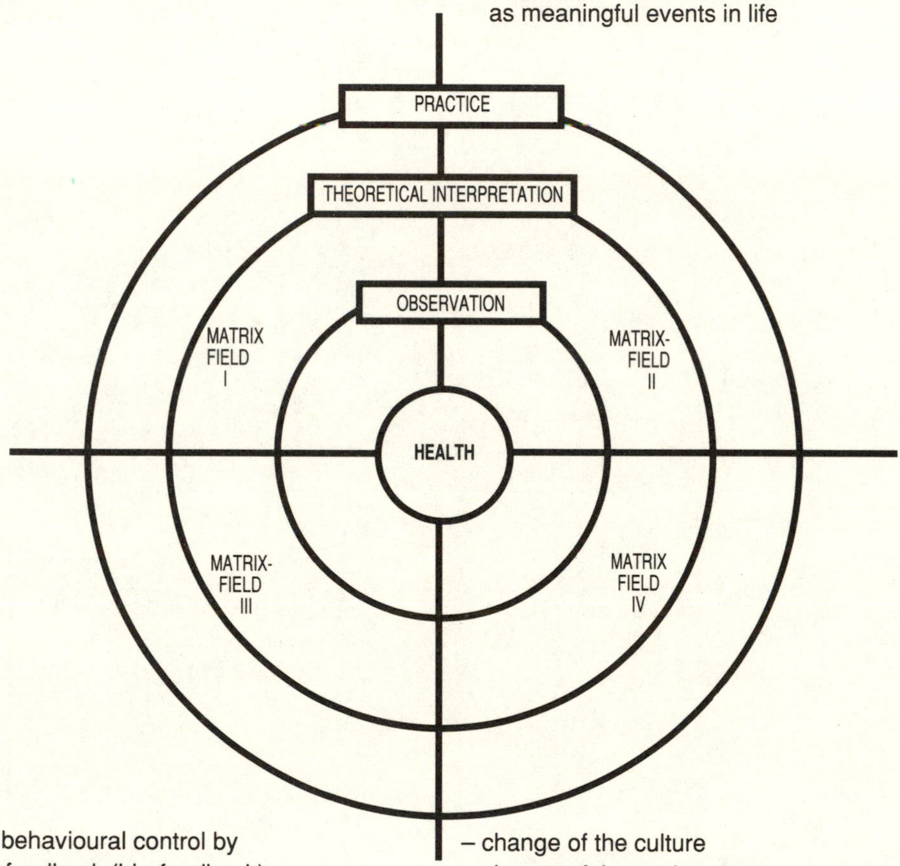

- behavioural control by
  feedback (bio-feedback)
- co-operation between all
  parts of the system
- individual health is a
  reflection of the health of
  the whole system

- change of the culture
- change of the environment
- change of social policy
- change of social relationships between people

Figure 1.4 Relations between paradigms, observations and health practices

pitfalls. One pitfall is the dream of a 'Unified Science'. This idea refers to the supposed possibility of integrating all theories in one single theory or meta-theory. It is connected to a deep, unfulfilled desire, more spiritual than scientific, to know nature. Many great philosophers, writers and scientists dreamed of the unity of knowledge.[5] This is an abstract utopia, not a concrete one. Another pitfall is the concept that this unified theory would represent the truth about reality. This idea is unrealistic, mythical and impossible to realize. It denies the dialectics between science and the observed reality. There is also the pitfall of denying that it is possible to reach some integration on meta-levels. This would consolidate the existence of an endless series of *ad hoc* theories. What we need is theoretical integration in a middle area, between unified theories on the one hand, and ad hoc theories on the other hand. It is an ongoing process. There will never be a last, final point. But at the same time, the attempt at integration could bring us further on our way to a more profound understanding of health.

(b) We also need to develop completely new paradigms and theories. A major opportunity is the creation of a bridge between current scientific knowledge and the personal and existential knowledge of everyday life. For such a model we can find ingredients in biographical literature and research, in body work and meditation, in the different self-awareness exercises of psycho-therapeutical schools, the self-help movement, and so on.

Such a model has to include the knowledge of 'lay' people and especially those who have had the opportunity to experience relevant situations (such as illness or long-lasting health, vitality or longevity). Thus the scientific knowledge of pain has to be linked with the experience of people in pain (Kaplun 1989; Vrancken 1989)! We need to institutionalize a dialogue on this subject and to include it in academic institutions in order to exchange information between those two groups. Health, health promotion and preventive health programmes refer first of all to a *praxis*. We need greater connections between a science of health and this praxis.

(c) Finally, I want to reflect on a social process which is a main context for our current discussion, because our discussion is part of this process. Diemer (1976–8) has described the development of worldviews in Western civilization through a typified process which contains different stages (ancient, medieval, etc.). The main development is in the difference between the symbiotic, medieval world and the contemporary disrupted and endlessly divided world. The world has changed and is further changing towards more and more complexity. Social relations become more and more fragmented and complex, and the identity of the individual is more a function of a place in a relative network instead of a stable relationship with the surrounding world. This evolution can be illustrated by the

19

increase of different health professions and the increase of the complexity of health institutions.

This trend of our civilization is stimulated by anonymous social forces. We cannot turn back to the Middle Ages, to a theological, symbiotic worldview. Yet we cannot deny the negative consequences of the exponential distortion of the worldview of today: the ecological catastrophe, the lack of meaning, the impossibility to control societal processes, etc. (Kickbusch 1989). There is a need for a new transition in social evolution. Not back to the past, but a move into the unknown future. This step needs to generate more meaning in society, more humanity in the world, more harmony and less distortion. This cannot be based on the old power structures and social control mechanisms. A fundamentalist reaction is not a real solution. The new worldview has to be built upon the free choice of every individual. As we don't have a model for it, so one has to be created. A first step is to start a dialogue about our dreams for the future: a dialogue which includes the whole, the past and the present, the inside and the outside, the objectivity and the subjectivity and most of all: the people of our planet. The new model has to be built on the unity of humanity as a whole.

REFERENCES

Bateson, G. (ed.) (1972) *Steps to an Ecology of Mind*, San Francisco: Chandler.
——— (1979) *Mind and Nature*, London: Fontana Paperbacks.
Beck, U. (1986) *Risikogesellschaft. Auf dem Weg in eine andere Moderne*, Frankfurt: Suhrkamp.
Berg, J. H. van den (1964) *Leven in meervoud. Een metabletisch onderzoek*, Nijkerk: G. F. Callenbach.
Bohm, D. (1983) *Wholeness and the Implicate Order*, London: Routledge & Kegan Paul.
Bourdieu, J.-P. (1977) *Outline of a Theory of Practice*, Cambridge/London/New York/Melbourne: Cambridge University Press.
Bourdieu, P., Chamboredon, J. C. and Passeron, J. C. (1968) *Le métier de sociologue, s.l.*
Callebaut, W. and Pinxten, R. (1987) *Evolutionary Epistemology. A Multiparadigm Program*, Dordrecht: D. Reidel.
Campbell, J. (1988) *The Power of the Myth*, New York: Doubleday.
Capra, F. (1975) *The Tao of Physics*, Berkeley: Shambhala.
——— (1982) *The Turning Point*, New York: Simon & Schuster.
——— (1986) 'Wholeness and health', *Holistic Medicine* 1: 145–59.
——— (1988) *Uncommon Wisdom: Conversations with Remarkable People*, New York: Simon & Schuster.
Compernolle, Th. (1981) 'De zelfbehandeling van recidiverende angstaanvallen en andere hyperventilatieverschijnselen', *Zelfhulptechnieken* (Self-Care Techniques), loose-leaf edition, vol. 2, Deventer/Antwerp: Van Loghum Slaterus.
Cox, H. (1978) *Oostwaarts. De belofte en bedreiging van de neo-oosterse bewegingen*, Baarn/Utrecht: Ambo.
Crawford, R. (1980) 'Healthism and the medicalisation of everyday life', *International Journal of Health Services* 10(3): 365–88.

Diemer, A. (1976–8) *Elementarkurs Philosophie. Philosophische Antropologie*, Düsseldorf: Econ.

Dijksterhuis, E. (1989) *De mechanisering van het Wereldbeeld. De geschiedenis van het natuurwetenschappelijk denken*, Amsterdam: Meulenhoff.

Dixhoorn, J. van (1990) *Relaxation Therapy in Cardiac Rehabilitation. A Randomised Controlled Clinical Trial of Breathing Awareness as a Relaxation Method in the Rehabilitation after Myocardial Infarction*, diss., Rotterdam: Erasmusuniversiteit.

Elias, N. (1969) *Uber den Prozess der Zivilisation. Soziogenetische und psychogenetische Untersuchungen*, 2 vols, Bern and München: Francke Verlag.

—— (1990) *Ueber sich selbst*, Frankfurt am Main: Suhrkamp.

Europäische Monographien zur Forschung in Gesundheitserziehung (1984) nr 6, Wien: Bundesministerium für Gesundheit and Umwelt Schuz.

Fortmann, H. (1971) *Inleiding tot de cultuurpsychologie*, Deel I, Bilthoven: Ambo.

Foucault, M. (1971) *Madness and Civilization: a History of Insanity in the Age of Reason*, London: Routledge.

—— (1973) *The Birth of the Clinic: an Archeology of Medical Perception*, New York: Vintage Books.

—— (1976) *Mental Illness and Psychology*, New York: Harper.

Gleichmann, P., Goudsblom, J. and Korte, H. (1977) *Macht und Zivilization. Materialien zu Norbert Elias' Zivilizationstheorie*, Part I, Frankfurt am Main: Surhrkamp.

—— (1984) *Macht und Zivilization. Materialien zu Norbert Elias' Zivilizationstheorie*, Part II, Frankfurt am Main: Surhrkamp.

Göpel, E. (1987a) *Lebensmodelle und ihre methodischen Konsequenzen für die Gesundheitsbildung*, Materialien des Oberstufen-Kollegs, Bielefeld: Oberstufen-Kollegs, Universität Bielefeld.

—— (1987b) *Beiträge zur Diskussion*, Materialien des Oberstufen-Kollegs, Bielefeld: Oberstufen-Kollegs, Universität Bielefeld.

Goudsblom, J. (1977) 'Civilisatie, besmettingsangst en hygiëne. Beschouwingen over een aspekt van het Europese civilisatieproces', *Amsterdams Sociologisch Tijdschrift* 271–300.

Hampen-Turner, Ch. (1981) *Maps of the Mind. Charts and Concepts of the Mind and its Labyrinths*, New York: Macmillan.

Helman, C. (1984) 'Culture, health and illnesses. An introduction for health professionals', Bristol/London/Boston: Wright.

Hesse, H. (1972) *Das Glasperlenspiel*, Küsnacht: Surhkamp.

Hoefnagel, A. H. J. M. (1977) 'Systeembenadering en sociologie', in Rademaker and Bergman (eds.), *Sociologische stromingen*, Utrecht: Het Spectrum/Intermediair.

Hurrelmann, K. (1988) *Sozialisation und Gesundheit. Somatische, psychische und soziale Risikofaktoren im Lebenslauf*, Weinheim/München: Juventa Verlag.

Ingrosso, M. (ed.) (1987) *Della prevenzione della malattia alla promozione della salute*, Milan: Franco Angeli.

Jantsch, E. (1980) *The Self-Organizing Universe: Scientific and Human Implications of the Emerging Paradigm of Evolution*, Oxford: Pergamon.

Kaplun, A. (ed.) (1989) *Health Promotion and Chronic Illness. Discovering a New Quality of Health*, Cologne: Federal Centre for Health Education.

Keeney, B. (1983) *Aesthetics of Change*, New York: The Guilford Press.

Kickbusch, I. (1989) *Good Planets are Hard to Find*, WHO Healthy Cities Papers, no. 5, Copenhagen: Fadl.

Kleinman, A. (1980) *Patients and Healers in the Context of Culture*, Berkeley: University of California Press.

Lafaille, R. (1984) 'Self-help as self-care', in S. Hatch and I. Kickbusch (eds), *Self-Help and Health*, Copenhagen: WHO, pp. 169–76.

—— (1991) *Programma's voor Gezonde Leefwijze. Een aanzet tot een culturologische en synthetische analyse* (Healthy Life-Style Programmes. A Contribution to the Development of a Culturological and Synthetic Analysis), Antwerp: International Institute for Advanced Health Studies.

Lafaille, R., Debaene, L. and Lebeer, J. (1989) 'The Center of Health Science at the University of Antwerp', *The International Journal of Health Promotion* 4(4): 349–53.

Lafaille, R., Geelen, K., van Aalderen, H., Cuvilier, F., van Dijk, P., Janssens, H., Lambooij-Clabbers, E. and Severne, L. (eds) (1981–5) *Zelfhulptechnieken* (Self-Care Techniques), loose-leaf edition, 3 vols, Deventer/Antwerp: Van Loghum Slaterus.

Lafaille, R. and Hiemstra, H. (1990) 'The Regimen of Salerno, a contemporary analysis', *Health Promotion International* 5(1): 57–74.

Lafaille, R. and Lebeer, J. (1991) 'The relevance of life histories for understanding health and healing', *Advances* 7(4): 16–31.

Lévi-Strauss, C. (1949) *Les Structures Elémentaires de la parenté* (The Elementary Structures of Kinship), Paris: PUF.

—— (1952) *Anthropologie Structurale* (Structural Anthropology), Paris: Plon.

McKeown, Th. (1979) *The Role of Medicine. Dream, Mirage or Nemesis?*, Oxford: Basil Blackwell.

McNeill, H. (1976) *Plagues and Peoples*, New York: Doubleday.

Maturana, H. R. and Varela, F. J. (1988) *The Tree of Knowledge. The Biological Roots of Human Understanding*, Boston and London: Shambhala.

Metz, W. (1975) *Pijn: een teer punt. Een fundamenteel geneeskundig onderzoek*, Nijkerk: G. F. Callenbach.

Niehoff, J.-U. and Wolters, P. (1990) *Ernährung und Prävention. Körpergewichte – ein Beispiel präventionstheoretischer Probleme*, Berlin: Wissenschaftszentrum Berlin für Sozialforschung.

Oud, J. H. L. (1978) *Systeem-methodologie in sociaal-wetenschappelijk onderzoek*, Nijmegen: Alfa.

Parsons, T. (1951), *The Social System*, New York: Free Press of Glencoe.

—— (1977) *Social Systems and the Evolution of Action Theory*, New York: Free Press of Glencoe.

—— (1978) *Action Theory and the Human Condition*, New York: Free Press of Glencoe.

—— (1979/1980) 'On theory and metatheory', *Humboldt Journal of Social Relations* 7(1).

Pauling, L. (1987) 'Een goede Voeding voor een Gezond Leven', *Chaos* 9: 28–35.

Peeters, H. F. M. (1978) *Historische gedragswetenshap. Een bijdrage tot de studie van menselijk gedrag op de lange termijn*, Meppel/Amsterdam: Boom.

Prigogine, I. and Stengers, I. (1983) *Order out of Chaos*, New York: Bantam Books.

Robes, J. (1988) *De gezonde burger. De gezondheid als norm* (The Healthy Citizen), Nijmegen: SUN.

Roszak, T. (1976) *Unfinished Animal*, New York: Harper & Row.

Sartre, J.-P. (1962) *De Heilige Genet (Saint Genet, comédien et martyr)*, Utrecht: Bijleveld.

Schipperges, H. (1985) *Homo Patiens. Zur Geschichte der kranken Menschen*, München/Zürich: Piper.

Swaan, A. de (1978) *Uitgaansbeperking en uitgaansangst; over de verschuiving van bevelshuishouding naar onderhandelingshuishouding*, Inaugural Address, Amsterdam: Boom.

Thom, F. (1972) *Stabilité structurelle et morphogénèse*, New York: Benjamin.

Verbrugh, H. S. (1978) *Paradigma's en begripsontwikkeling in de ziekteleer*, Haarlem: De Toorts.

—— (1983) *Nieuw besef van ziekte en ziek zijn. Over veranderingen in het mensbeeld van de medische wetenschap*, Haarlem: De Toorts.

Vrancken, A. M. E. (1989) *Chronische Pijn, het Kruis van de Geneeskunde*, Alblasserdam: Haveka.

Vries, M. de (1985) *Het Behoud van Leven* (The Preservation of Life), Utrecht: Bohn, Scheltema & Holkema.

Wilber, K. (1981a) *No Boundary*, Boulder: Shambhala.

—— (1981b) *Up From Eden. A Transpersonal View of Human Evolution*, London: Routledge & Kegan Paul.

—— (1983) *Eye to Eye: the Quest for the New Paradigm*, Garden City, N.Y.: Anchor Books.

Wildiers, M. (1988) *Kosmologie in de Westerse cultuur*, Kapellen: DNB/Uitgeverij Pelckmans.

Zeeman, E. C. (1977) *Catastrophe Theory. Selected Papers 1972–1977*, London: Addison-Wesley.

Zijderveld, A. C. (1966) *Institutionalisering*, Hilversum/Antwerp: Paul Brand.

Part II

THE HISTORICAL PERSPECTIVE

2

HUMAN HEALTH AND PHILOSOPHIES OF LIFE[1]

Eberhard Göpel

1 CONCEPTS OF HEALTH AND HUMAN INSECURITY

Health and illness are polar, complementary concepts, which depict those factors that endanger both life itself and our safe views on life. In this latter sense the concept of illness refers to specific instances of suffering which call for an explanation and understanding. It is supposed that if they can be understood within a cultural explanatory framework, their disquieting effect will be diminished. The concept of health, on the other hand, characterizes a complementary attempt, through the interpretation of the processes of illness and suffering, to gain and use insights with which to improve the life chances of those so far unaffected by illness.

In all cultures, the basic insecurity of human existence due to the spread of disease, pain, or impending death has led to the development of systems of healing that undertake to explain and treat the inexplicable, and in this way to limit fear and despair.

The activity of healers and therapists, whose function is to mediate between the concerned individual and his or her social environment, promises relief from the immediacy of potentially threatening events. (They are relied upon to offer assurance when the explanatory patterns of those affected fail, giving way to fear and hopelessness. This legitimation largely depends on their actual competence to provide effective help, and requires that the explanation and interpretation offered is convincing to the individual concerned, thereby increasing trust in the effectiveness of treatment and providing additional support through this psychological function.)

In the European tradition, the development of a modern medical system of healing was an addition to, and later a demarcation from (earlier systems of healing which were religious in origin.) It began with a view of nature which presupposed that biological processes followed their own laws and could be described and influenced according to knowable rules and methods. This view considered disease to be a malfunction of biological mechanisms which could be recognized and classified as such and manipulated by physical or chemical intervention. This biomedical model of health

27

and illness, which has become a monopoly, has replaced other interpretations of human suffering in our society. Society has transferred to the medical profession the exclusive right to define what is illness, who is ill or healthy, and what is to happen to those who are ill. The adoption of the biomedical model has made it possible to adapt the treatment of human suffering to the functional requirements of industrial society. The rise of the 'medical–industrial complex' to one of the biggest single economic sectors of society, accounting for over 10 per cent of the gross national product, is, however, becoming 'counterproductive' (Illich 1975). The medicalization of society, beyond this critical level, does more harm than good.

The reductionist perspective of the biomedical model of illness is proving increasingly inadequate to cope with the globalization of life risks and their emotional impact. It has begun to wane, and as it does so a variety of new and formerly repressed patterns of explanation, offering a more appropriate approach to human suffering, are making their (re)appearance. At the beginning of industrial society it was the material need that made people associate and develop a system of 'social security'; in the emergent 'risk society' (Beck 1986) the emphasis is going to be on socially mediated insecurities which will lead people to new types of collective action.

A historical evaluation might help us to focus more sharply on these changes in the concepts of health and illness and to differentiate between the basic patterns of the thought that give rise to them. The following questions will be addressed by this contribution:

- In which way has the history of thought dealt with health, disease, and death in European history?
- What kind of connections to the ecological and social environments are crucial for the preservation of health and the prevention of illness?
- What principles and life models are used to explain the dynamics of health, disease, and death?
- What is the the role of the individual?
- What disadvantages and dangers are inherent in fixed and dogmatic views of life and concepts of health?
- What are emancipatory approaches to life and health?

2 THE HISTORICAL DEVELOPMENT OF GENERALIZATIONS ABOUT LIFE AND ITS RISKS

2.1 The otherworld and the quest for meaning

(In most primitive societies, to be ill signifies mortal danger and could even be felt as an attack by the souls of the dead.)In the archaic consciousness no distinction is made between human being and environment, between the animate and inanimate worlds (In those days people felt surrounded by

28

mysterious forces which decided over life and death, and in order to reduce their fears they attempted to give meaning to what they observed. Everything was considered to possess a living soul in the way that humans experienced themselves: animals, plants, rocks, lakes, rivers, etc. Everything that moved and changed was seen as an expression of an inexplicable force that existed as true reality, albeit invisible, and determined the life of the living. To feel weak and sick in this context was highly significant. It could mean preparing for death. The undividedness of sensory experience, in which dreams, illusions, and fantasies became intermingled, in combination with pain, suffering, disease, and death, revealed the full weight of insecurity, which affected other members of the tribe. Relief in such a situation could be obtained by way of establishing a projective distance: the naming of the other, the opposite, the otherworld, the invisible.

In archaic–animistic cultures all natural phenomena are considered potential influences on life's manifestations which have to be placated by way of rituals. Animistic views explain disease as something that enters the healthy person from outside; there are invisible spirits lurking everywhere to harm human beings and animals, wreaking illness on people, preventing fertility, causing impotence, or even driving people insane and possessing them.

In a further stage of cultural development illness is usually not thought to be the product of sheer malice, but associated with some higher objective pursued by the gods, such as rewarding the good and punishing the bad. The main cause for illness is thought to be the violation of divine laws, prescriptions, and taboos. In this context, illness would be a punishment for offending the gods and demons and the demonstration of a reigning, but invisible power. The personification of a demon or an avenging god thus becomes an important precondition for the development of a correct ritual in which the demon is invoked, if he or she is to be exorcized or placated.

Accordingly, ideas concerning prevention and treatment of illness are based on the prevailing animistic–demonic view of illness. Commonly, illness can be avoided by submission, imploring the gods for mercy, making sacrifices such as food, drink, or valuables to placate the wrathful spirits, or by attempting to lure away, exorcize, trick, overpower, or banish and conjure the evil demon that causes disease.

All such attempts to control disease are greatly enhanced if they are undertaken by medicine men, shamans, or priests who have a special relationship with spirits, demons, and gods, perfected by specific rituals and trance states. Under their direction and with the aid of auxiliary spirits, healing gods, holy men, and specific ceremonies, attempts are made to strengthen the power and faith of the living and overcome malfunctions and temptations. Illness is not seen as an isolated event affecting one individual, but is held always to threaten the entire social system and communal life. Shamanistic therapy therefore addresses both the patient and the healthy individuals and attempts to create a new balance within nature, in human relations, and in the relations with the invisible realm of the spirits. The effectiveness of such strategies against suffering therefore requires a basic consensus: on the faith of the therapist in his or her techniques, on the faith of the patient in the power of ritual, and on the faith of the group in the validity of these efforts. As a form of preventive medicine and promotion of health, such a method is bound up with a closed community of the type that has been partly preserved in traditionally agrarian societies or in sects and religious groups.

In tribal religions the ideal situation had always been one of untroubled relationship between the collective and the gods inhabiting the cosmic order. In the monotheistic religions, however, which emerge in the advanced feudal civilizations of antiquity, such as China, India, and the Near East, this view underwent a radical change. Human beings no longer felt part of a primarily intact relationship with the invisible or numinous that confronted them in the form of a personal god or divine principle. Instead, burdened with original sin and blindness (maya) from the beginning, they are doomed to a fateful existence that separates them from salvation. In such a hopeless situation, ideas of 'salvation' and 'divine inspiration' and the myth of a rebirth of human beings on a higher level, become increasingly important. The conception of health and illness is embedded in a comprehensive, unified doctrine of a divine principle, of the nature of human beings and their destiny in the world.

The healing objective, or cure, in terms of these conceptions of life is primarily religious in character and, accordingly, the remedies are initially those of religious worship, i.e. usually sacramental in kind. Medical remedies of natural origin are also admitted, but only as additional means of aiding the Creator's Plan, as God's hands, so to speak. According to God's intentions illness is no coincidence; rather, through it He reveals His unfathomable will. Illness would thus make sense within the context of His inscrutable ways. It may also be meted out as punishment for the sins a person has committed, the consequence, perhaps, of a free, but blasphemous act. However, God may also pursue other ends. Illness may even be a revelation of divine grace, an expression of God's love, a reward bestowed upon a person. God imposes suffering on people as an ordeal and for purgation, providing them with an opportunity for doing good deeds which will count on Judgement Day. Attaching little value to his or her weak, mortal body, a sick person may go some way toward securing eternal life.

In Christian teachings concerning physical life, the body is considered to be merely the mortal, transient frame of the soul. In this sense, the important thing is to ensure physical health by first ensuring spiritual health. Accordingly, to be free of illness is not of primary concern, since suffering may speed the extinction of guilt. Personal acceptance of illness and atonement are therefore preconditions for any cure. The salvation of the soul enjoys absolute primacy over the preservation of the body.

While in the Roman Catholic tradition we witness a somewhat passive attitude to the body, the latter being regarded merely as the mortal frame of the soul, the Protestant tradition sees the emergence of a rigid morality opposed to worldly pleasures. This aims at controlling the body, and demands that all its sensory impulses be subject to a moral imperative. The Protestant ethic with its cardinal virtues of hard work, sense of duty, frugality, and modesty has emancipated itself from the mysteries of body and soul, and turned the body into an instrument of worldly labour in the pursuit of the soul's salvation. Physical and spiritual health will be granted to those who attempt, through personal endeavour, to contribute their share to fulfilling the Creator's Plan and the moral prescriptions.

While in the Roman Catholic Church the primary mission is seen in the redemption of the human soul, in the Protestant–Pietistic tradition there emerges a secular charitable movement as manifested, for instance, by the work of the Salvation Army which cares for the human soul, but also provides soap, hot soup, and a place to sleep. The belief was that a human individual, if left alone in times of danger to body and soul, would be unable to cope with weakness and temptation. What was needed was care and encouragement by a wider supportive group or social community.

The development of religious forms and contents more or less emulates and reflects the structures of social development. Thus the emergence of stable feudal structures was accompanied by corresponding ecclesiastical–religious hierarchies, just as in modern time, the development of bourgeois commerce and trade relations was promoted by a corresponding Protestant ethic.

Generally, it can be stated that the need for a transcendental meaning increases to the degree that the meaning people attach to, or are able to realize in, their everyday lives decreases. Similarly, the need for mediation through intermediaries in the guise of shamans, magicians, priests, doctors, gurus, and 'health educators' is the greater, the less people are actively and responsibly involved in shaping the framework of their everyday lives.

2.2 Nature and the quest for order

2.2.1 The magic order

Magic is the attempt to identify, and systematically exploit, correspondences and sympathies between natural phenomena, often attempting to use them for healing purposes. Within the magic world view, human beings can conceive of themselves as being one with the manifestations of their environment and fellow living beings, albeit for reasons that remain obscure to them. The natural world is inextricably bound up with their personal being. In the magic world, relations to the world are emotional rather than rational,

31

and affective expressions, like sympathy and antipathy, are vitally important for the relationship between human beings and objects. In keeping with the principle of singularity, recommendations for remedies to cure illness frequently name unusual animals, plants, fruits, stones, or metals, which are thought to possess special powers solely by virtue of their peculiar structure. Another magic motif is the principle of similitude, i.e. the faith in the power of sameness or similarity. As a rule, similarity attaches to the surface of things – e.g. it is a similarity between symptoms (such as swellings, discolorations, deformations, and sensations), or between diseased organs, and similarly shaped, formed, or coloured structures encountered in nature and the human environment. Remedies can be chosen because of the strength of such identities.

In the endeavour to discover the laws guiding life's manifestations and to exploit them, people watched the stars with a view to their potential influence on the good or bad fortune of people, and on phenomena such as famines, epidemics, illness, and death. In the Orient, thousands of years ago, the faith in the power of the stars led to the development of very advanced astrological systems. The influence of cosmic rhythms on people's lives was accepted as self-evident. The sleeping and waking routine of all creatures followed the course of the sun and adhered to a day-and-night rhythm. With the coming of daylight and warmth, nature was seen to awake. Births occurred more frequently in the early hours, as did deaths. The rising and setting of the moon corresponded to the tidal ebb and flow. The female cycle more or less followed the monthly cycle. The role of annual cycles, too, clearly affected the manifestations of life. In the spring plants ended their winter period of rest, started to grow and blossom in a certain sequential order. Quite obviously nature followed the cosmic laws.

Within the context of faith in a divine predestination of the world, the stellar constellation at a person's birth was regarded as being of great importance for that person's fate, and provided the basis for horoscopes, divination, and prophecies with the aid of astrological calculations.

2.2.2 The hierarchical order

Gradually the mythical interpretation of the world leads to a more systematic observation of nature and, coupled with rational conclusions, to a (re)construction of reality. The account of the origin and course of the universe restricts itself to observable substances and processes and dispenses with the idea of divine intervention or Providence. Faith is superseded by reason which strives to replace traditional religion with methodological thought. As an animate link in the divine cosmic system, the human individual occupies a central position and should have the ability to use his or her reason as a tool to structure the relations with fellow humans as well as nature, since the intellectual faculty of reason is given to all human beings.

For this reason, the systematic observation of nature, of its laws and ordering principles is held to be vitally important for answering some fundamental questions confronting humanity, such as:

- What is a human being?
- How are human beings constructed considering their complexity?
- What is their reciprocal relationship with the cosmos?
- Which laws and principles underlie the functional systems in a healthy or sick person?

Contrary to human assumptions which can err, natural laws are considered unalterable points of reference and hence irrefutable.

In the effort to construct a symbolic representation of the basic framework of life and all its aspects, the following basic patterns can be clearly distinguished.

2.2.2.1 The quest for symmetries and equivalencies

The ancient view presupposes that human beings and nature are part of a cosmic harmony which is divinely inspired and therefore exhibits an order based on specific rules. Medicine, or healing, draws on practical experience with the natural and cosmic laws and becomes nature's handmaiden and an agent of health by undertaking to restore the disturbed harmony. In this process, the physician is merely an auxiliary to the natural force – that universal operating principle whose aim is the elimination of defects and the harmonization of dissonances. Elements, humours, and temperaments are manifestations of a natural harmony whose balance is constantly at risk. This balance has to be corrected, if necessary, by a life conducted with greater awareness (for example in terms of diet). The physician, who is familiar with the principles involved, is supposed to take on the role of companion and advisor to the patient, and even of philosophically trained expert, in all phases and crises that occur in a person's biography. From this worldview developed the 'schema of basic elements' with its many differentiations.

This schema starts from the idea that the basic elements of the cosmic order are formed by the four substances air, water, fire, and earth and their interplay, and operate in human and all nature. Accordingly, any treatment aimed to restore the necessary equilibrium of the elements, works with the two pairs of opposing qualities which define them, such as hot and cold, humid and dry. Galen, in particular, introduced this concept into European medicine, where its influence became very widespread and could be felt until modern times.

In the tradition of the Galenic system of medical practice, however, this ancient view of life, which originally was philosophically oriented, increasingly turned into a dogma of prescription and became a closed medical doctrine which, well into the modern age, largely determined medical theory.

2.2.2.2 Hierarchical orders

The notion of a hierarchical structuring of nature had already been elaborated by Aristotle and was absorbed into the Christian tradition. The Aristotelian view of nature holds that everything strives for the highest level of perfection, although this is only achieved with varying degrees of success. A kind of 'ladder', or hierarchy, of nature can thus be constructed, ranging from minerals, plants, and the higher animal species to human beings, at the top rung of the ladder. Perfection is conceived of as exemplifying superiority and unchangingness. Living beings are physical and mortal, above them are the physical and eternal celestial bodies. At an even higher level we find the rational soul, which is immaterial and eternal. The supreme position is held by God, the most unchanging and therefore most concrete of all substances, Who realizes His possibilities in the most perfect manner. Ambition and love can only proceed in an upward direction, looking to a higher level of perfection. The conception of different degrees of perfection constitutes the natural order of social relations: as the slave looks to his master, so the wife looks to her husband, and a human being to God. In such a theocentric world order the worth of human beings derives from their relative closeness to God. The Neoplatonic view of the universe is a typical example of the adaptation of different elements taken from various worldviews to the framework of the Christian tradition.

By dividing the world into elemental, celestial, and intellectual spheres, the lower always being controlled by the higher sphere and absorbing the influence of the latter's forces, it is essential, in keeping with God's creation, to establish a universal order through an appropriate equilibrium of forces. Since the human individual is the pivot and purpose of creation, God, through the celestial forces and their manifestations in the shape of rocks, herbs, metals, etc., communicates to, and places at the disposal of, the human individual the specific powers and shapes of those objects. The faith in their effectiveness thus rests on the interrelatedness of all things with the original cause as well as on the relationship with those divine models and eternal principles. Everything has its particular place within the original order from where it has sprung and receives its vitality, and the force of the herbs, rocks, metals, animals, words, prayers, and all that exists is a divine gift. God's principle manifests itself in the cosmic forces present in our world.

2.2.3 The rhythmic development

The Chinese tradition, in particular, has developed a system of nature which focuses on the growth and decay of life's manifestations. Accordingly, the descriptive patterns refer mostly to cyclic, rhythmic, or spiral objects. One example from the Asian tradition is the spiral of life.

According to this system, mankind occupies the terminal point of a gigantic spiral of life which once emerged from the ocean of the one infinity. The animal kingdom, in which mankind represents the last manifestation, is part of the plant kingdom, on which it depends directly or indirectly for life and nourishment. There are no distinct boundaries between the two, since plant life is constantly changing or being transformed into parts of animal life. A continuous rotation of the spiral initially proceeds from within the spiral itself. The universal spiral of life proceeds in seven stages of development along an inward path. Within these developmental stages human beings with their psychic and intellectual attributes represent the most sophisticated form of life. With the arrival of human beings and other highly developed animal species the centripetal process of the spiral reaches its climax: the inward process of physical and material manifestation culminates in anthropogenesis, whence, starting at the centre of the spiral, it flows in the opposite direction, returning to preceding manifestations. As the spiral expands, disintegration and spiritualization ensue, dissolving personal and individual identities, and the process of life merges with the one infinity, the origin of everything.

2.3 The body and the quest for function

2.3.1 Models of mechanical functioning

In the Renaissance the physical reality offered new qualities of experience. Technical skills, and questions of how best to solve practical problems, are transformed into a theoretical knowledge of nature and its characteristics. Around the beginning of the seventeenth century and in the wake of the mutual exchange of experience between technicians, artists, craftsmen, and theorists, a conviction emerged that human creativity did not so much outwit nature, as join it in harmony. The introduction of a mathematical theory of proportionalities and the application of geometry to technical problems did not exclude the human body. Leonardo da Vinci (1452–1519), for example, tried to depict the human being and nature in aesthetically pleasing, but also in technically concrete terms. For the first time, an unconditional study of nature included a study of the human body, which signified a radical break with the medieval view of the world.

The human individual no longer appears as a part of nature which becomes a passive mechanical system, alien to the freedom and function assigned to the human mind. With the aid of experimental tools the human mind may gain access to a vantage point from where God too looks upon the world, and to that divine plan whose perceptible expression is the world. In pushing nature to lower levels, everything that is not part of her becomes glorified: God and the human being. Having cut the umbilical cord that tied human beings to nature, they could then damage, rape, and exploit her without feeling pain or shame, since they only damaged an external object, and no longer a subject with which they were once connected.

In the Aristotelian–Scholastic feudal period human beings were still an organic constituent of nature of which they felt part, and which they tried to observe in order to know more about her. In modern times the human individual, as a subject, confronts nature as an object and, by way of appropriate experiments using measuring instruments, tries to identify her laws and, in the process, constructs a natural science dissociated and defined in abstraction from nature as it originally existed. The natural laws, i.e. the laws underlying and determining the phenomenon, could only be discovered because nature was turned into an object.

As far as the scientific study of human diseases was concerned, this new approach meant that researchers came to restrict their observation to increasingly smaller segments of human life processes in order to reach towards the ideal case, which was the physical–mathematical description of those processes (see Figure 2.1).

Causes of disease

1. In the disturbed equilibrium of the humours (holistic model in antiquity).
2. In the organs (Morgagni 1761).
3. In the tissues (Bichat 1802).
4. In the cell (Virchow 1858).
5. In the DNA molecule (Miescher 1869, Watson and Crick 1963).

(Quoted from Feyerabend 1963: 242)

Figure 2.1 Causes of disease according to modern science

First and foremost it was the philosophy of Descartes that established the radical distinction between the cognizing subject and the passive object of cognition. Descartes started by designing a new theory of substances which replaced the Aristotelian coexistence of the physical and the spiritual, with a radical body–mind dualism. According to this doctrine, the body as *res extensa* (the 'extended thing') differs substantively from the soul as *res cogitans* (the 'cognizing thing') such that an influence from the mental–spiritual sphere upon the bodily sphere or vice versa becomes unthinkable.

For Descartes the existence of God remained an important element in his scientific philosophy. In subsequent centuries, however, scholars tended to avoid any reference to God, and developed their theories on the basis of the

Cartesian distinction. The humanities concentrated on *res cogitans* and the natural sciences on *res extensa*. The strict division between the mental, subjective sphere and the natural, objective sphere has until today remained the basis of the European scientific tradition and has prevented (e.g. in medicine) a unified form of description of the manifestions of human life. In his efforts to develop a perfect natural science, Descartes extended his mechanistic view of matter to include living organisms. Plants and animals were machines to him. Human beings were inhabited by a soul endowed with reason, the soul being linked with the body via the pineal gland in the centre of the brain. The human body, however, was no more than an animalistic machine. In order to prove that living organisms were just automatons, Descartes gave a detailed description of how the movements and various biological functions of the body could be reduced to mechanical processes. This particular view of the living organism has had a decisive impact on the search for a science of life. The detailed description of mechanisms that, put together, produce living organisms came to be the dominant method in biology and medicine. The Cartesian method proved particularly succesful in biology, but at the same time limited the orientation of scientific research. The problem was that *the success achieved in treating living organisms as machines eventually made the scientists believe they actually were machines.*

2.3.2 Models of dynamic functioning

Apart from the attempt undertaken by the mechanistic–materialist world view to explain life's manifestations as phenomena of the external, material world which can be perceived by the senses, those involved with understanding health have continued to seek an understanding of these phenomena by looking to the interior world, which is experienced consciously and emotionally. Many now believe that it is not just mechanically operative factors that determine life processes, but that there are 'forces' at work in every living thing which defy being subsumed under physicochemical laws. In such psychodynamic models of life, the relationship of the immaterial, 'soul-like' forces to the material physical substrate was often conceived of in such a way that the former represented the animating, i.e. actuating and guiding, principle, while the material substance formed the passively controlled substrate.

Psychodynamic conceptions of medicine have made a renewed impact since the end of the nineteenth century. They imply that the roots both of organic diseases and mental disorders are to be found in psychic factors and influences, and lead to appropriate therapies or theories of healing. According to this view, either the body is the main arena in which psychodynamic illness occurs (psychosomatics), or psychodynamics manifests itself predominantly in the arena of mental or emotional symptoms in the form of psychic disorder or mental derangement. It is claimed that, in principle, a major role, both in causing an illness and curing it, must be attributed to the psychic domain, and that this applies to all physical and psychological (mental) disorders.

In the eighteenth century bio-dynamistic models of medicine had become increasingly important, and were opposed to an animistic metaphysics of the soul, on the one hand, and to plain mechanical concepts, on the other. The faith in invisible 'forces' had been fuelled by the demonstrable force of gravitation and magnetism as well as the peculiar attractive and repulsive forces of electrical bodies. The ideas of a *vis vitalis* that developed in the eighteenth century followed a distinct course of their own, in two ways: that the machine theory of organic phenomena (reproduction, growth, nutrition, regulatory functions, self-movement, etc.), was not adequate, but neither was the reliance on the metaphysics of the soul in view of the contemporary ideal of scientific knowledge. Generally, ideas of *vis vitalis* focused on problems of growth or development and the organization of living beings, and provided the hypothetical framework for phenomena that had yet to be sufficiently explained in mechanistic terms (e.g. the phenomenon of the stimulation and excitation of muscle fibres). The *vis vitalis* was seen as a sustaining force that is continuously active in all processes of life and, as a formative force to build the organism. This force constitutes the life element of the body, protecting it against destruction and decay, and in case of illness, especially in inflammatory processes, plays a regenerative role. The *vis vitalis* endows the organism and its parts with an ability to respond to stimuli. Death means the loss of *vis vitalis*. Illness is nothing but a reaction of the vital forces to the stimulus of illness. For this reason, the latter will persist until the *vis vitalis* has overcome the cause, or until the cause has overcome the *vis vitalis*. Thus, the expression of the symptoms of an illness depends on the *vis vitalis*, its strength or weakness, its quality, its distribution and presence in the organs and tissue, its dependence on the constitution, life-style, nutrition, light, air, and state of mind – all of these factors determine the level of excitability and the reaction to pathogenetic factors. In the history of biology and medicine the assumption of a *vis vitalis*, or energy as a formative principle of biological organisms, has produced numerous diverse theories. Some examples are Paracelsus's iatrochemistry, Mesmer's magnetic therapy, Abraham's radionics, Steiner's ethereal theory, Hahnemann's homeopathy, and Reich's organic theory.

2.3.3 *The iatro-technical concept of clinical medicine*

In the history of modern philosophy and natural science vitalist conceptions have repeatedly emerged as a reaction to the dominant mechanistic concept of life. Since mechanistic explanations of life were held to be inadequate, the belief arose that life could only be explained and understood by resorting to some independent life force. The controversy over vitalism and mechanism, or idealism and materialism, was to become fundamental for the history of European philosophy, biology, and medicine. The assumption of 'higher principles' by the vitalists was countered with 'deeper principles' proposed by the mechanists. The search for deeper principles became increasingly focused on the body's inside. Morgagni's theory of organic pathology, which attributed the cause of disease to a single organ, was superseded by Bichat's histological pathology, according to which disorders in the body tissue necessarily led to malfunctions of an organ. In the middle of the last century, a further step in the localization of pathological processes was taken by Rudolf Virchow and his cellular pathology. He was able to show that, rather than organs or tissue, it was individual cells that were the carriers of pathological processes. He found that all diseases could ultimately be traced back to active or passive disorders in the body's cells whose performance changed depending on the state of their molecular structure, which could in turn be ascertained by cytopathological tests. The place of classical humoral pathology, which had emphasized the importance of bodily humours, was now taken by a comprehensive system of *Solidarpathologie*, a comprehensive theory of the localization of diseases. Virchow's idea of localizability provided the foundation for modern surgery since, on the basis of this pathology, it now made sense to remove a diseased organ surgically or deal with its malfunction by surgical means. Cellular pathology, advanced by recent developments in molecular pathology and genetic engineering, has now led to a situation where medicine has abandoned the holistic view of the human body and become ever more fragmented in outlook, focusing on individual parts such as organs, tissues, cells, and molecular structures (see Figure 2.1). The inevitable specialization of doctors has only accelerated this process. Cellular pathology had far-reaching implications for the medical profession: where there is no sick body, but only sick cells, there is no need to treat the whole body. A local pathology obviously requires a local therapy. The rest of the body, of the individual, is reduced to being the material carrier of the disease.

In the middle of the last century a dominant natural science concept of the organism emerged as well as a new theory of a medicine modelled on natural science and technology. Owing to the monopolization of medical training at university clinics the iatro-technical concept of medicine was granted the status of 'traditional medicine' which, in turn, became the generally accepted framework of state-funded medical research and teaching.

From among the diversity of health experts found in the European tradition during that period a monopoly was thus secured by the academically trained medical experts, and since then medical science, i.e. traditional medicine, has functioned as the only legitimate, publicly entrusted 'guardian of health', authorized to give directives regarding behaviour to the individual and to establish norms which, directly or indirectly, become binding on society as a whole.

For the 'non-expert' – i.e. the patient as much as anyone else who, within the framework of public consciousness, is in need of orientation – the concentration of expert knowledge and decision-making power in the domain of academic medicine carries an important advantage for the individual: namely, of being relieved from having to take his or her own decisions. This is linked to the conviction that the given, and accepted, decision is the best possible in terms of the modern criterion of rationality. In this way, the choice of a specific behaviour can be left to the experts, while the non-experts accept this voluntary dependence because it appears to them to be freely chosen and rational.

The public debate on the self-interests of the medical 'clergy' and the limitations of the biomedical creed, have however, made today's public sceptical and shaken their confidence in the expert power of the medical profession. With the decline of the monopoly of definition held by medical 'expertocrats' a new marketplace of possibilities has opened up. In addition to a revival of some older, seemingly dependable, or at least less toxic remedies and cures, and the consultation of other, more trustworthy experts, the emphasis is clearly shifting to self-interest and self-reliance, to empowering of the individual.

2.4 The subject and the search for a programme

The formulation of the classical laws of mechanics by Newton had significant repercussions on the worldview of bourgeois society. John Locke, a contemporary philosopher and political scientist, argued for example that all of mankind's problems could be solved if only the laws of mechanics were to be applied. He based his philosophy on an 'atomistic' structure of society: every human being is isolated from the other in the sense that he or she has no 'organic' relationship to the former; all human beings in a society are individuals, mass points, so to speak, whose relations are regulated via the forces and counterforces that operate among them. Political power has to ensure that these forces and counterforces are balanced and reach a stable equilibrium.

While, in the feudal system, society was considered an organism in which everyone occupied a clearly ascribed position with prescribed functions, in the bourgeois system of society everyone is a free individual.

Just as in the feudal system the ruling class legitimized their system of government by invoking Aristotelianism, so the bourgeois society supported their specific form of rule by referring to the new science of mechanics. The system of individualism defined the relational framework of society and was then transferred to natural science and medicine itself. Both deal with separate 'individual' facts, theories and principles, or diseases and symptoms, while any overarching ideas are banished to the realm of metaphysics.

The disintegrative method of the natural sciences, which increasingly fragmentized the domains of human experience, was compensated by the development of a philosophy which, by reviving concepts from antiquity, gradually took over the task of religion: to provide a unified framework for our knowledge of the world.

A survey of popular models in psychology and medicine (model of the machine, model of the organism, model of the self) will help to illustrate the enormous impact of metaphoric models, derived from non-medical or psychological contexts, on the perception and constitution of reality.

2.4.1 *The model of the machine*

The basic pattern of this model is the machine metaphor: the human being is viewed as a machine, with more concrete definitions emerging in parallel with technological developments.

In psychological terms, the machine model defines the human individual in terms of an apparatus activated and directed by forces outside its control. Moreover, the passiveness of the machine, its dependence on external act-ivation, and the absence of an autonomous level of self-control, in human terms, imply an individual as having no aims or objectives of his or her own. Instead, this model turns the individual into a mere recipient of externally fixed goals. Since human existence is dependent on the environment, a human being has to be made like a machine in order that he or she be able to live and act. For example, if human behaviour is considered determined by specific genetic factors, a direct intervention in the determining genetic structure, viz. genetic engineering, appears to be the appropriate method to remove behavioural disorders.

Another approach of this model is the social conditioning of human behaviour, as developed by the behaviourists. If both nature and the social environment are assumed to determine human behaviour, then such be-haviour is simply the combined influence of genetic features plus social conditioning. This is the view of psychobiology.

The approximation of machine and life processes has proceeded along two entirely different paths. On the one hand, machines have been constructed that more or less exactly simulate processes of life. The second approach is an attempt to gain access to, and control of, the structure of life as it already exists; to initiate biological processes under human control. This began in agriculture, farming and stockbreeding, and is currently carried further by experiments involving the manipulation of genes and the artificial pro-duction of life.

Machine and man, artificial product and living being – both seem to be moving closer to one another. Artificial objects now tend to exhibit more of those characteristics that some decades ago seemed reserved for human beings. Computer technology, in particular, endows machines with a capacity for largely autonomous action and flexible responses as well as 'abstract' thinking. There have been spectacular achievements in the technologization of life, especially in medicine, biology, and biotechnology. Products consisting of living and dead matter seem to have a promising future: spare parts for human beings, bio-prostheses, bio-computers, etc. The boundaries between machines and living beings are becoming blurred.

Hopes or fears that artificially constructed machines might acquire properties of living beings or even enter a symbiosis with life, thereby turning into some sort of hybrid monsters, part machine and part human being, are unfounded. Even if two such vastly different systems were to be joined, the combination would remain superficial: an artificial heart implanted in the human body remains a mechanical pump, and the rest of the body continues with its own live functions. Instead, the interaction of human being and machine is always envisaged in bodily terms. For in the tradition of European thought, mind and body are considered separate entities, and the sense of self is thought to be seated in the mind. In this sense, no problem is involved in exchanging the heart because this is simply a matter for the body which does not affect mental identity. If human identity is to be found in the mind, the body must necessarily be a kind of life-support system for the mind. In this view there is no difference in principle between the body and an artificial machine. The living heart can be replaced with an artificial one.

This argument can easily be taken further: in comparing living and artificial systems, or machines, it turns out that, in various respects, the machine comes out better. It is more 'perfect' in many ways, and thus becomes a kind of 'role model' for human beings. The affinity between human being and machine is not found on the bodily, but on the mental level. It owes its existence to a specific way of reasoning, a radical abstraction from the living human body with its capacity for concrete sensuous experience.

2.4.2 *The model of the organism*

This model describes a holistic view of the human being. The organism is considered to be an organized totality, a 'system', which develops as part of an interactive process with its environment. The constituent elements of the organism are not significant in themselves, but are related to the whole entity. In psychological terms this means, for instance, that a particular behaviour makes sense only within the specific context in which it has to fulfil its constitutive function, and that an isolated description of behaviour therefore makes 'no sense'. Starting from the observation that in biology there are no constant forms independent of metabolic processes, the model of the organism abandons the separation between *structure* and *function* in favour of a dynamic process. The process is the basic principle, which can be understood in terms of functions, and is ultimately oriented to maintaining the organismic structure.

Adaptation is therefore a central concept of this model: the active adaptation to environmental conditions as a permanent function of the organism. Part of this adaptation is developmental, as the organism grows towards a state of optimum functioning. Contrary to earlier metaphysical ideas of a predetermination of life, this model does not regard development as preprogrammed.

Since organism and environment have to be viewed as a unit, change and development cannot be explained solely in terms of some external influence or internal determinant. Rather, it is the mutual interactions that define the organism. For example, experience and learning may influence development, but cannot be the sole cause, for the organism is not constituted by environmental influences, but possesses an original structure that regulates its exchange with the environment. For this reason, experience and learning have to be viewed against a background of the constructive activity of the organism.

2.4.3 The model of the self

Contrary to the machine concept, which presupposes a passive human being, the model of the self implies an active individual who, in contrast to the model of the organism, does not act involuntarily and unconsciously, but is guided by a purpose and by intentions, corresponding to his or her subjective interpretation. Here, the responsibility for self-directed action rests on the belief that human beings are capable of recognizing and controlling their motives, interests, and intentions.

For an understanding of human action an analysis of external causes and events is inadequate. Rather, it is necessary to try to reconstruct subjective reasons. Only a hermeneutical understanding can ultimately yield a comprehensive explanation of the 'meaning' of a specific act, and the actor's intention. If we formulate the reasons for an act in this way, we rationalize it, i.e. we define it as being meaningful and reasonable.

The model of the 'homo economicus' starts from the assumption that human beings are basically guided by self-interests and the greatest possible utility that might accrue to them. Within a framework of biologically determined 'drives', such as eating, drinking, sleeping, sexuality, aggression, etc., human beings need to set themselves a rational daily routine which serves as a basis for an adequate 'economy of drives'. This extended 'economic system of life', characterizes the bourgeois health ideal with its implicit goal of an orderly way of life. Those who upset or even ignore it altogether can be said to act irrationally.

Since personal characteristics frequently disturb social interaction, it appears sensible for the individual, in interaction with others, to anticipate the expectations of such a meaningful environment and fulfil a complementary role. In this model society is felt to exert a constraining influence and is perceived as a powerful and objectionable element to which individuals have to submit. In terms of role theory the 'real and authentic' individual is found 'behind' society, in a private domain, at home.

3 AN OUTLOOK ON POSTMODERNISM

The historical review has shown, I believe, that the profound change in consciousness that came with the beginning of modern time has deeply affected the way people, at least in Europe, interpret their lives.

The orientations guiding the individual's relationship with his or her social and ecological environments have undergone a progressive shift towards self-control. In this process the constraints imposed on life are internalized. As this happened, man broadened his decision-making and achieved a subjective freedom of choice. He replaced the religious ethic anchored in faith with a universalistic, rationalized secular morality.

It was a central element of the early bourgeois political views to create a community of free, equal, and autonomous individuals, and to establish rules for such a community based on a negotiation of interests and the application of natural laws. The task of medicine in this context was to develop rules for conducting a life based on the laws of nature and to eliminate metaphysical interpretations of life from the public domain.

Health and security became socially desirable conditions, once people saw themselves as autonomous, acting individuals. The term 'security', and the social value it carries in this connection, denotes a popular desire, and generally accepted norm, that everyone should partake in a (secure) future offering certainty and absence of anxiety. 'Health' refers to the sense of security relating to the physical and mental processes of life. External social order and security ensures internal security to the extent that the external order reflects internal needs. A continuation of the status quo appears as the primary concern – the preservation of health and protection against untimely death.

From the perspective of health these principles imply the following preventive concepts:

(a) A whole world: continuity of social development and stability of traditional communities based on a binding value system in which basic personal needs are safeguarded.

(b) Total planning: the guaranteeing of possible future(s) within the framework of a totally controlled process of development by means of institutional structures designed to safeguard individual well-being.

(c) Self-control: a continuous supervision of one's own normality as well as control and discipline of one's physical and mental responses through personal effort towards self-chosen goals.

These three models continue to figure as important fictions in the current debate on health policy issues. They are mechanistic, and so are largely responsible for the alienation in people's attitudes to their own suffering and that of others. The political programmes tied up with these models ensure a systematic barrier to dealing with health and disease in a more humane way.

The dynamics of the modern economy has built a process of social differentiation, and this fragments a unified contextual framework of subject and world. People feel a loss of coherence and waning of orientation, and are forced to be autonomously acting subjects – whether they want to be 'free' in this sense or not. The mirror image of this process of individualization is a feeling of existential isolation and the need for an individually defined meaning of life.

The constant invoking of the self that we are witnessing at present, such as the encouragements to self-help, may hide individual isolation, involving anxiety and psychological stress – especially in view of the fact that this campaign questions the individual's loyalties to the powers that be. The health issue is an example of the brinkmanship of modern consciousness: modern man needs a self-determined realization of personal potential, while, at the same time, retaining both freedom and security.

In the current transition to 'self'-control and the declining power of science-based medicine, older patterns of explanation are being reactivated and enjoy renewed authority. Against this background, the task of the health sciences might conceivably be the development of a metatheory capable of integrating the collective wisdom of times past. Failing this we will continue to face a situation in which marketable fragments of knowledge, holding out promises of security and meaningfulness, will be offered for consumption to an existentially troubled public.

REFERENCES

Abholz, H. H. (1982) *Risikofaktorenmedizin*, Berlin.
Achenbach, G. B. (1985) *Das Prinzip Heilung*, Köln.
Alexander, F. (1921) *Psychosomatische Medizin*, Berlin.
Antonovsky, A. (1987) *Unravelling the Mystery of Health*, London: Jossey-Bass.
Arguelles, J. M. (1974) *Das grosse Mandala-Buch*, Freiburg.
Argument-Jahrbücher für kritische Medizin (1982–6) Bd. 7, 8, 9, 10, 11, West Berlin.
Attali, J. (1979) *Von der Magie zur Computer-Medizin*, Frankfurt: Campus.
Bammé, A., Feuerstein, X., Genth, R., Holling, E., Kuhle, R. and Kempien, P. (1983) *Maschinen-Menschen, Mensch-Maschinen. Grundrisse einer sozialen Beziehung*, Reinbek: Rowohlt.
Bateson, G. (ed.) (1972) *Steps to an Ecology of Mind*, San Francisco: Chandler.
—— (1979) *Mind and Nature. A Necessary Unit*, London: Fontana Paperbacks.
Beck, U. (1986) *Risikogesellschaft. Auf dem Weg in eine andere Moderne*, Frankfurt: Suhnkamp.

EBERHARD GÖPEL

Berkeley Holistic Health Center (1982) *Das Buch der ganzheitlichten Gesundheit*, Bern.

Berr, M. A. (1984) *Die Sprache des Körpers. Wider den Vandalismus des Rationalen*, Frankfurt.

Bloch, J. R. and Maier, W. (eds) *Wachstum der Grenzen. Selbstorganisation in der Natur und die Zukunft der Gesellschaft*, Frankfurt.

Capra, F. (1982) *The Turning Point*, New York: Simon & Schuster.

Clark, R. and Pinchuck, T. (1984) *Medizin für Anfänger*, Reinbek bei Hamburg.

Diemer, A. (1978) *Elementarkurs Philosophie. Philosophische Anthropologie*, Düsseldorf: Econ.

Ehrenreich, J. (ed.) (1978) *The Cultural Crisis of Modern Medicine*, London: Monthly Review Press.

Elias, N. (1977) *Über den Prozess der Zivilisation*, 2 vols, Frankfurt.

Erikson, E. H. (1979) *Identität und Lebenszyklus*, Frankfurt.

Feyerabend, P. (eds) (1983) *Wissenschaft und Tradition*, Zürich.

Foss, L. and Rothenberg, K. (1987) *The Second Medical Revolution. From Biomedicine to Infomedicine*, Boston and London: Shambhala Publications.

Foucault, M. (1976) *Die Geburt der Klinik. Eine Archäologie des klinischen Blicks*, Frankfurt.

Franke, E. (1986) *Sport und Gesundheit*, Reinbek.

Freidson, E. (1979) *Profession of Medicine*, Chicago: Dodd, Mead & Company.

Funkkolleg (1982) *Umwelt und Gesundheit-Aspekte einer sozialen Medizin*, 2 vols, Frankfurt.

Gehrke, C. (eds) (1981) *Ich habe einen Körper*, München.

Geissler, B. and Thoma, P. (1975) *Medizinsociologie. Einführung in ihre Gesundheitsbegriffe und Probleme*, Frankfurt.

Grossinger, R. (1980) *Planet Medicine*, New York: Anchor Press/Doubleday.

—— (1982) *Wege des Heilens. Von Schamanismus der Steinzeit zur heutigen alternativen Medizin*, München.

Hamden-Turner, C. (1982) *Modelle des Menschen. Ein Handbuch des menschlichen Bewusstseins*, Weinheim.

Hartmann, F. (1984) *Patient, Artz und Medizin*, Göttingen.

Haug, W. F. (1986) *Faschisierung des bürgerlichen Subjektes. Die Ideologie der gesunden Normalität und die Ausrottungspraktiken im deutschen Faschismus*, Argument-Sonderband nr 80, Berlin.

Herzog, W. (1984) *Modell und Theorie in der Psychologie*, Göttingen.

Hoffmann, D. (1984) *Leibes-Übung. Ein Streitbuch über die neuen Medien in der Körperkultur*, Darmstadt.

Holzkamp, K. (1983) *Grundlegung der Psychologie*, Frankfurt.

Illich, I. (1975) *Medical Nemesis*, New York: Pantheon.

—— (1976) *Limits to Medicine*, London: Marion Boyars.

Imhof, A. E. (eds) (1983) *Der Mensch und sein Körper. Von der Antike bis heute*, München.

Jäckle, R. (1985) *Gegen den Mythos ganzheitlicher Medizin*, Hamburg.

Jacob, W. (1978) *Kranksein und Krankheit. Anthropologische Grundlagen einer Theorie der Medizin*, Heidelberg.

Jantsch, E. (1982) *Die Selbsorganisation des Universums*, München.

Juchli, L. (1985a) *Einführung in die Krankenpflege*, Stuttgart, 11.A.

—— (1985b) *Heilen durch Wiederentdecken der Ganzheit*, Zürich.

Kamper, D. and Gutanding, F. (1982) *Selbstkontrolle. Dokumente zur Geschichte einer Obsession*, Marburg.

Kamper, D. and Rittner, V. (1976) *Zur Geschichte des Körpers. Perspektiven der Anthropologie*, München.

Kamper, D. and Wulf, Ch. (1982) Die Wiederkehr des Körpers, Frankfurt.

Kapp, K. W. (1983) *Erneuerung der Sozialwissenschaften. Ein Versuch zur Integration und Humanisierung*, Frankfurt.

Kleinman, A. (1980) *Patients and Healers in the Context of Culture*, London: University of California Press.

Krusche, M. (ed.) (1982) *Ökologisches Bauen*, Berlin: Deutsche Bau-Verlag.

Kursbuch 82 (1985) *Die Therapiegesellschaft*, Berlin.

Kushi, M. (1979) *Das Buch der Makrobiotik*, Frankfurt.

Lasch, C. (1982) *Das Zeitalter des Narzissmus*, München.

Leontjew, A. N. (1979) *Tätigkeit, Bewusstsein, Persönlichkeit*, Köln.

Lerche, D. (1980) *Du streichelst mich nie!*, Frankfurt.

—— (1984) *Keiner versteht mich!*, Frankfurt.

Lippe, R. zur (1978) *Am eigenen Leibe. Zur ökonomie des Lebens*, Frankfurt.

Lüth, P. (1986) *Das Ende der Medizin? Entdeckung der neuen Gesundheit*, Stuttgart.

Lutz, R. (1983) *Bewusstseins (R)evolution*, Ökolog-Buch 2, Weinheim.

—— (1984) *Frauen-Zünfte. Ganzheitliche feministische Ansätze, Erfahrungen und Lebenskonzepte*, Ökolog-Buch 3, Weinheim.

Maturana, H. R. (1982) *Erkennen: Die Organisation der Wirklichkeit*, Braunschweig.

Mies, M. (1980) 'Gesellschafliche Ursprünge der geschlechtlichen Arbeitsteilung', *Beiträge zur feministischen Theorie und Praxis* M.3.

Milz, H. (1985) *Ganzheitliche Medizin. Neue Wege zur Gesundheit*, Frankfurt.

Müller, R. W. (1977) *Geld und Geist – Zur Entstehungsgeschichte von Identitätsbewusstsein und Rationalität seit der Antike*, Frankfurt.

Oerter, R., Dreher, E. and Dreher, M. (1977) *Kognitive Sozialisation und subjektive Struktur*, München.

Oppl, H. and Weber-Falkensammler, H. (eds) *Soziale Arbeit im Gesundheitswesen*, Frankfurt.

Petzold, H. (ed.) (1982) *Methodenintegration in der Psychotherapie*, Paderborn.

—— (1985) *Leiblichkeit. Philosophische, gesellschaftliche und therapeutische Perspektiven*, Paderborn.

Prigogine, J. and Stengers, J. (1981) *Dialog mit der Natur. Neue Wege naturwissenschaftlichen Denkens*, München.

Psychologie heute (1981) Sonderband 'Lebenswandel. Die Veränderung des Alltags', Weinheim.

—— (1985) Sonderband 'Die Körper, die wir sind. Mit Leib und Seele leben', Weinheim.

Rexelius, G. and Grubitzsch, S. (1981) *Psychologie. Theorien-Methoden-Arbeitsfelder*, Reinbek.

Richter, H. E. (1979) *Der Gotteskomplex*, Reinbek.

Rosenberg, A. (1976) *Kreuzmeditation*, München.

Rosenbrock, R. and Hauss, F. (1985) *Krankenkassen und Prävention*, Berlin.

Rössner, H. (1986) *Der ganze Mensch. Aspekte einer pragmatischen Anthropologie*, München.

Rothschuh, K. E. (1963) *Theorien des Organismus*, München.

—— (1965) *Prinzipien der Medizin*, München.

—— (1978) *Konzepte der Medizin in Vergangenheit und Gegenwart*, Stuttgart: Hippokrates.

Sabetti, S. (1985) *Lebensenergie*, Bern.

Salmon, J. W. (ed.) *Alternative Medicines. Popular and Policy Perspectives*, London: Tavistock.

Schaefer, H. and Blohmke, M. (1976) *Sozialmedizin*, Stuttgart.

Schipperges, H. (1970) *Moderne Medizin im Spiegel der Geschichte*, Stuttgart.
—— (1985) *Homo patiens. Zur Geschichte des kranken Menschen*, München.
Schmid, H. D. (1986) *Grundriss der Persönlichkeitspsychologie*, Frankfurt.
Schmidt, H. (1983) *Didaktik des Ethikunterrichts*, Stuttgart, Bd 1.
Sirananda Yoga Zentrum (1985) *Yoga. Für alle Lebensstugen in Bildern*, München.
Varela, F. (1985) 'Der kreative Zirkel. Skissen zur Naturgeschichte der Rückbezüglichkeit', in Watzlawick, P., *Die erfundene Wirklichkeit*, München.
Wambach, M. M. (ed.) (1983) *Der Mensch als Risiko*, Frankfurt.
Wenzel, E. (1986) *Die Ökologie des Körpers*, Frankfurt.
Wilber, K. (ed.) (1986) *Das holographische Weltbild*, Berlin.
Willi, J. (1985) *Koevolution. Die Kunst gemeinsamen Wachsens*, Reinbek.

Part III

DEVELOPMENTS IN DIFFERENT SCIENTIFIC FIELDS

3

FRONTIERS OF BIOLOGY IN RELATION TO HEALTH

Brian Goodwin

1 INTRODUCTION

The Peckham Experiment was one of the first attempts in Britain to identify the social, communal, and family conditions that allow for the expression of good health in individuals. The initial study, carried out in London in the 1920s and 1930s, articulated a new view of the science of health that continues to be highly relevant today. A fervent proponent of this view, Dr Kenneth Barlow (1988), describes the definition of health that emerged in the following terms.

In the individual there is seen to be growth and development – such development displaying remarkable differentiation. Consideration revealed that in the family there was likewise a potential for growth and development, capable of influencing the members of the family. Over the fifteen years or so during which the lives of member families were under intermittent review, it became clear that a further potential resided in the community to which the experiment had given rise.

In each of these three instances, the development of the individual, the development of a family, and the development of a community, there is seen to be a potential. When that potential is realised there is health. Health can accordingly be recognised as the realisation of this biological potential. In each case structures are created and it is the use of these structures, the function to which they are put, which allows of judgement of the excellence or otherwise – the degree of health – which has been achieved. But commonly what is potential is not achieved; what is possible is seen not to have occurred. The result in such circumstances is not health – but ill health. It is ill health which involves doctors. It is the high incidence of ill health which has led public knowledge to confuse health, which when realised stands in no need of the repairs which doctors can offer – with ill health. The point to be made is that some conditions facilitate the realisation of potential and health whilst others are inhibitory.

51

The stress here is quite clearly on *development*, the dynamic process of growth and transformation that all living systems, individuals and communities, undergo in realizing their potential. Whatever positively contributes to that process will enhance health; whatever inhibits it will result in ill-health.

This perspective significantly shifts the focus of health care in a particular direction, emphasizing the dynamic integrity of the person and the community, because growth, development, and transformation can be understood only within the context of the whole, not the part. Of course if a part is diseased and is threatening the integrity of the individual, something must be done about it. Western medicine has developed extremely effective ways of treating damaged or diseased parts. The fractured or crushed limb is expertly set or amputated; infection is effectively treated with antibiotics. But the Western analytical tradition has not been so successful at managing the whole body, at maintaining the balance of energies and activities that facilitate growth, development, and transformation.

This is where the so-called complementary therapies are having their greatest influence, filling a deficiency in traditional medical practice. People have been turning to herbal medicine, to homoeopathy, to acupuncture, to reflexology, to artistic and other therapies. They seek a restoration of dynamic harmony after being stressed into imbalance or locked into physical or psychological postures that prevent the normal process of growth and development. It is interesting to take a systematic look at the relationship of these complementary practices to conventional medicine, to locate the various traditions on a spectrum of their qualities and relate these to the qualities of the person.

An informative way of examining the relationships of the different healing and health-care traditions is in terms of a spectrum that extends from the mechanical at one extreme to the symbolic or cognitive at the other (Figure 3.1). In between, a representative range of the therapies are ordered in terms of progressively more subtle and coherent energy fields, passing through chemical, physical and emotional to the realm of the fully conscious being where a person's fundamental system of values and life priorities are involved. I owe this scheme to a colleague, Dr Michael Evans of the Anthroposophical Medical Association. What jumps out at us is that the specialized branches of the Western healing tradition have tended to focus on the two extremes of the spectrum, the mechanical/chemical and the cognitive. This corresponds to the dualistic Cartesian split in our culture between matter and mind. A vast range of levels in between (of which there are many more than those shown in the figure) correspond to all the secondary qualities that were left out of the science that emerged in the seventeenth century. They tend to be ignored except for the best traditions of general practice. We have paid dearly for this split, this dualism, but we are moving now towards wholeness and the gaps are being filled by therapies appropriate to the different levels of our being. The spectrum constitutes a

MECHANICAL–CHEMICAL – PHYSICAL (ENERGY FIELDS) – EMOTION–FEELING – COGNITION

Surgery   Cognitive psychotherapy

Conventional medicine   General counselling

Osteopathy   Spiritual counselling

Herbal medicine   Hypnotherapy

Massage   Depth psychotherapies

Homoeopathic medicine   Meditation techniques

Acupuncture   Artistic therapies

Biofeedback   Eurhythmy therapy

Relaxation techniques

Figure 3.1 The spectrum of the therapies

whole: conventional and complementary traditions belong together as a unity. Wilber (1987) also describes an integrative scheme that unites the therapies by identifying them with different levels of the human being, between which there should be no boundaries.

A developmental perspective on health and healing carries with it certain emphases. One of these is a quality of experience and orientation that is a characteristic of good practice in all the therapies. This stresses wholeness, balance, harmony and treats the whole person in social context. But since health is a dynamic process of development – the realization of potential – it is natural that this harmony be interrupted by periods of transformation. Now it is during the episodes of transformation, while the individual is changing from one state of dynamic balance to another, that he or she is most vulnerable, most open. This is when minor illness is most likely to occur: dis-ease, lack of harmony, the transitional state necessary for change to occur. So disease may be diagnostic of transformation, an indicator of change. Hence it need not be something to be avoided and prevented at all costs, but can provide a physical symptom of a transition in the continual process of maturation and individuation. Childhood diseases have this quality. Adult illness can be read in a similar manner: a symptom of change, or a need to change. If this need to transform and develop is prevented as a result of the life-style, social, political or economic circumstance or ecological conditions, then serious illness can result. Similarly, particular states of disease may be diagnostic of particular conditions in society.

I am by profession a developmental biologist, so this perspective on life as growth, development and transformation is literally bread and butter to me.

Some of the most dramatic images of change from one state of order to another come from the observation of natural transitions in developing organisms. The metamorphosis of the plant meristem from leaf to flower, of the pupa into the butterfly, of the tadpole into the frog, are stunning transformations, all involving a form of dis-ease – the arrest or dissolution of certain structures, tissues and cells and the growth of others. From the perspective of the dying cell, the part, metamorphosis is a tragedy; from the viewpoint of the whole organism, it is what life is about. Recognizing the dynamic of the whole is what makes sense of the change in its parts.

2   THE BIOLOGICAL FOUNDATIONS OF HEALTH

The emphasis on wholeness and transformation in the healing arts brings me to another theme. What is the research strategy that is appropriate to a science of health whose objective is the husbandry of the whole organism in a state of dynamic harmony?

Medical research is grounded in biology, as it should be. But biological research has been dominated by certain assumptions, certain paradigms that impose a very specific form on the subject. The whole organism as a real entity has actually disappeared from contemporary biology, that is, from Neo-Darwinism. This may seem to be an improbable assertion, since the primary entity that biologists deal with and which is the object of biological study is the organism, whether in the context of evolutionary, developmental, physiological, biochemical, or other area of investigation. However, it is a fact that in biology as currently taught there is no theory of the organism as a self-organizing, dynamic, transforming entity, always undergoing developmental and evolutionary change, whether it be rapid or slow. There are excellent theories of inheritance, of molecular composition, and of many aspects of physiological process from the point of view of regulation and control (homoeostasis). But nowhere in mainstream contemporary biology do we find a theory of the whole organism. This is why research in both biology and medicine concentrates on parts – on anatomy, physiological systems, the immune system and its constituent cells and molecules, the genetics of disease, etc. This is all excellent knowledge as far as it goes. But it fails at the level of the whole, because it inverts the correct order of priorities. In a machine, the whole is assembled from its parts. But in the organism, the parts are generated by the whole. We see this most clearly in embryonic development: the organism starts as a single, spherical egg, with no organs or tissues or differentiated cells. Then gradually, systematically, through episodes of slow and sudden change, the complex, integrated, multi-faceted being emerges with a wide range of organized, coherent levels of structure and function. It is this being, this organism, that medicine serves. But it cannot perform this service if there is no theory of organisms as integrated wholes. Since biology fails to deliver this crucial vision, medicine is impoverished and does what it can with the parts that are understood. And complementary therapies come in to fill this need. But the research base of a holistic biology must be developed to serve holistic medicine. And it is in fact growing rapidly. There is a major new movement in biology that confronts and seeks to achieve this objective, with an emerging research programme (Ho and Saunders 1984; Ho and Fox 1988; Goodwin, Sibatani and Webster 1989; Goodwin and Saunders 1989). How is it to be conducted? How do we investigate *wholes*?

A basic concept which provides the key to understanding organisms as integrated, transformational dynamic systems comes out of embryological studies. Parts of embryos have the capacity to develop into wholes. For example, if the two cells that result from the first cell division of a fertilized egg are separated, each one gives rise to a perfectly normal whole organism, even though in normal development each cell would have produced only half of the whole. This is how identical human twins are produced, by the accidental separation of the first two cells after fertilization of the ovum. The German embryologist Hans Driesch, who conducted such experiments with sea-urchin embryos, described the phenomenon as similar to the separation of a bar magnet into two parts (Driesch 1892). Before the separation there is a single magnetic field, with north and south poles. If the magnet is cut in two, separating the two poles, each one becomes a complete magnetic field, reconstituting the whole from a part.

This metaphor of a field has now become a well articulated theory in which these regulative properties of developing organisms are described in terms of the fundamental concept of the developmental field. This is defined as a domain of relational order in which the state of any part is determined by the states of other parts so that the whole has a characteristic property of integrity or completion. (See, e.g. Goodwin 1985, 1990). Organisms are generated by the action of these fields, which gives them their integral or holistic properties. The field property of organisms is also the basis of health and healing: it is what provides the intrinsic dynamic for a return to the state of wholeness and harmony after disturbance, whether the disturbance is physiological or physical, as in wound healing or liver regeneration. The regenerative capacities of adults varies greatly between species. Whereas simple organisms such as *Hydra* can regenerate the whole from small parts, and newts can regenerate whole limbs, human beings can regenerate only the tips of their fingers. This occurs only within the last section of the finger, distal to the terminal joint. And it is important that the wound be left open so that a wound epithelium forms, across which an electrical current flows. If the wound is closed by sewing a flap of mature skin over it, no current flows and no regeneration occurs. The surgical instinct to patch up the wound is in this case counter-productive. Overall, the fundamental property of organisms that underlies the self-healing process is the tendency of the developmental field to restore and maintain a condition of completeness and

of balance, of wholeness and harmony in the organism and in its relationships with the environment. (For an extended account of these properties,
see Goodwin 1992.)

The concept of the field or domain of relational order with a distributed
causality can also be used to describe the genesis of the person in a social
context (Ingold 1990). The developmental field of the organism is then
extended homologously to the realm of social relationships, in which a
similar type of coherent, ordering activity is at work that results in the
generation of the integral entities we call persons. These field concepts
provide a basis for understanding humans, both at the organismic and the
cultural levels, as dynamic, self-organizing, transforming wholes that are
constantly involved in a process of balanced change in which all parts are
engaged in maintaining a coherence that is simultaneously highly sensitive
and very robust. The human being is like a fountain whose form is always
restored after disturbance because there is a flow of all the components
through the system so that it can always regenerate its natural state of
dynamic balance. The human being is very sensitive to the environment
because there is a constant flow of environmental materials (nutrients) and
stimuli (information) through the organism; but the self-organizing field of
the organism attempts to maintain harmony. So we have the complementary
properties of extreme sensitivity and great robustness of the living state. The
study of organisms from this perspective provides a biological foundation
for the science of health.

3   NON-INVASIVE RESEARCH

There is also a research methodology that goes with the science of health.
In order to study wholes it is necessary to develop techniques that are
essentially *non-invasive*, so that the integrity of the organism is not compromised. Let me mention in this context a specific research programme that
I think is particularly significant in this respect. It is the study of spontaneous photon emission from living organisms and cells conducted by Dr
Fritz Popp, working in Kaiserslautern, West Germany. He has been observing and analysing the characteristics of photon emission across the visible
spectrum from plants, animals and cell culture. His remarkable deduction
from this work, based on carefully conducted analyses, is that the organism
is functioning somewhat like a multi-mode laser, since the emitted photons
have the characteristics diagnostic of a coherent source (Popp 1986; see also
Ho 1989). This raises an exciting possibility that coherent quantum-
mechanical states are fundamental to the dynamic order of living organisms.
This and related studies are clearly highly controversial, and rightly so, but
the possiblities revealed here are dramatic. One could imagine how photon
emission could be used for non-invasive monitoring that detects subtle
changes of body state in response to various types of change, dietary and

environmental, that seek to impove health, since what is being measured is a variable that correlates directly with a condition of internal coherence.

There is a great range of non-invasive diagnostic and therapeutic techniques that are now being developed to study and influence in subtle ways the distribution of energies and activities in the body, both in the conventional and the complementary areas. This reflects a development away from the mechanical and chemical interventions of conventional medicine, the left end of the spectrum of the therapies, towards more adequate ways of conducting research into the dynamic balance of the whole organism. In biology, a similar shift is resulting in a range of techniques whereby changes of state in organisms can be studied using *in vivo* signalling systems that do not interfere with normal activities and functions. These are the imaging techniques that provide space–time analyses of states of living organisms, providing the information necessary to understand the nature of the dynamic order and balance that underlies conditions of health and disease. The shift is away from the use of drugs and chemicals, and towards the use of normal metabolites and techniques that influence energies, primarily electromagnetic stimulation and monitoring. Among these are scanning devices such as those based on nuclear magnetic resource and SQUIDS, which measure very weak magnetic fields, and allow the imaging of electrical currents in the heart, the brain, or other part of the body while it is functioning normally. (Ioannides 1991; Swithenby 1987).

One of the big tasks ahead of us is the clean-up job of detoxifying our bodies and the environment. As reliance on chemical intervention wanes and the use of energy-balancing techniques increases, so the removal of extraneous chemicals from the body and the environment can proceed apace. It is widely recognized that health and the environment are the urgent, pressing challenges of today. The solution of these problems requires the same type of therapy for organism and environment, based on an understanding of the requirements of the whole. In relation to the planet, this vision has been articulated by James Lovelock in the Gaia hypothesis (Lovelock 1979), which sees the Earth as a living, developing organism, subject to similar processes of dynamic balance to those in our own bodies. So the way we heal ourselves, by detoxification and energetic balancing, is how we heal the planet. This is the health-care system of the future. It is a planetary, global network of action focused on the recognition that health is based on dynamic balance and development, a balance of resources within a continually transforming organism, community and culture. This engages humanity and all the other inhabitants of the planet in co-operative action.

REFERENCES

Barlow, K. (1988) *Recognizing Health*, Plymouth: Latimer Trend & Co.

Driesch, H. (1892) 'The potency of the first two cleavage cells in the development of echinoderms', in B. H. Willier and J. M. Oppenheimer (eds), *Foundations of Experimental Embryology*, London: Prentice-Hall.

Goodwin, B. C. (1985) 'What are the causes of morphogenesis?', *Bioessays* 3: 32–6.

—— (1990) 'Structuralism in biology', *Science Progress Oxford* 74: 227–44.

—— (1992) in Harman, W. (ed.) *Causality in Modern Science* (in press).

Goodwin, B. C. and Saunders, P. (eds) (1989) *Theoretical Biology: Epigenetic and Evolutionary Order from Complex Systems*, Edinburgh: Edinburgh University Press.

Goodwin, B. C., Sibatani, A. and Webster, G. C. (eds) (1989) *Dynamic Structures in Biology*, Edinburgh: Edinburgh University Press.

Ho, M.-W. (1989) 'Coherent excitations and the physical foundations of life', in B. Goodwin and P. Saunders (eds), *Theoretical Biology: Epigenetic and Evolutionary Order from Complex Systems*, Edinburgh: Edinburgh University Press, pp. 162–76.

Ho, M.-W. and Fox, S. W. (1988) (eds) *Evolutionary Processes and Metaphors*, New York: Wiley.

Ho, M.-W., and Saunders, P. T. (1984) *Beyond Neo-Darwinism: an Introduction to the New Evolutionary Paradigm*, London: Academic Press.

Ingold, T. (1990) 'An anthropologist looks at biology' (Curl Lecture, 1989), in *Man* (NS), 25: 208–29.

Ioannides, A. A. (1991) 'Brain function as revealed by current density analysis of MEG signals', in *Clinical Physics and Physiological Measurement* (in press).

Lovelock, J. (1979) *A New Look at Life on Earth*, Oxford: Oxford University Press.

Popp, F. A. (1986) 'On the coherence of ultraweak, photoemission from living tissues', in C. W. Kilminster (ed.), *Disequilibrium and Self-Organization*, Dordrecht: Reidel, pp. 207–30.

Swithenby, S. J. (1987) 'SQUID magnetometers: uses in medicine', *Phys. Technol.* 18: 17–24.

Wilber, K. (1987) *Without Boundaries*, Boulder: Shambala Press.

4

RELATIONS BETWEEN A SCIENCE OF HEALTH AND PHYSICS FROM MICROPHYSICS TO COSMOLOGY

Dirk K. Callebaut

1 INTRODUCTION*

In this article we investigate how physics and/or physicists can contribute to the development and foundation of a science of health. In some cases the relation is rather distant, for others it seems promising.

The plan of the article is as follows. First we will describe some general features of contemporary scientific development, seen by many as in the midst of a shift of paradigms. We will describe this development within physics and stress what is important from the point of view of the natural sciences. Needless to say, our treatment is selective and far from complete. With this information as a general background, we will then explore what can be learnt from the scientific tradition of physics for the further development of the health sciences. Thirdly, we will examine how physics can directly contribute to the solution of current health problems. And finally, we suggest how physics and macrophysics could be useful in discussing worldview, which is so important as the philosophical basis of a new science of health (see the chapter by Lafaille in this book).

2 A PARADIGM SHIFT IN PHYSICS?

The classical paradigm[1] of the natural sciences grew during the seventeenth to nineteenth century. It is usually symbolized by the pioneering discoveries of Isaac Newton. We will call this Newtonian paradigm the *classical model*.

The main features of this classical model are listed below. They are interrelated, and not all of them were fully embraced by all scientists in the indicated period:

- *Analytical logic.* Logical explanations (ultimately founded in formal logic and mathematics) are superior. Analytical procedures are normative for all sciences. This implies denial of meaning and purpose as

59

possible explanations for the occurrence of phenomena. This also implies a value-free, emotion-free attitude from scientists.

- *Empiricism*. The validity of theoretical constructions has to withstand experiment. An attempt must be made to control all confounding variables.
- *Determinism*. The universe is scientifically understood to be ultimately deterministic. Every phenomenon has its own cause(s).
- *Observationalism*. The observable world of objects and events is primary. All scientific knowledge is ultimately related to data obtained through the physical senses.
- *Mechanism*. The world is governed by universal, causal laws which can be discovered by science. Reality is a machine in which all parts are linked to each other in one causal chain. Natural phenomena are governed by a complex interplay of causes and effects.
- *Positivism*. It is assumed that the real world is what is physically measurable. There is an absolute preference for quantification. The universe is describable by facts or phenomena.
- *Reductionism*. Physical and chemical explanations have favoured status. It is assumed that we only come to a real understanding of phenomena through studying the behaviour of its constituent elements and that the universe consists, ultimately, of fundamental particles and their energetic relationships. The more elementary the particles, the more fundamental the explanation. Evolutionary processes (like ontogenesis, morphogenesis, regeneration, etc.) are explained by coded instruction in the genes or similar mechanisms.
- *Objectivism*. The truest information about objective reality is obtained by a completely independent, detached observer. Any influence of the observer is treated as a distortion of the observation.
- *Dualism*. A fundamental split exists between science and consciousness. Consciousness is treated at most as an epiphenomenon in the natural sciences. Many human scientists still believe that consciousness can be explained in a reductionistic manner (as the result of neurophysiological or biochemical processes in the brain). Altered states of consciousness are conceived as pathological states. The classical model is linked to an ontological assumption of separateness (Harman 1992) dividing the observer from his object, man from nature, mind from matter, scientific knowledge from human experience, science from society and science from religion.

The classical model is still very influential today. Scientific work is judged by reference to the basic assumptions of the classical model. The principles of the classical model are still generally accepted, although in a moderate manner. For example, experiments are no longer considered the exclusive road to proof in the human sciences. The dominance of the classical model has ambivalent effects. It has generated a vast amount of knowledge, both scientifical and practical. It has changed the world radically from the time of the industrial revolution. But at the same time, there is a heavy load of negative side-effects: the influence of the positivistic classical model has put human scientists on the wrong track and created the dominance of the biomedical model in the health sciences. The a-political, a-moral attitude of science, the separation of science and human experience, has led to severe ecological and social problems.

During the twentieth century many of the basic assumptions of the classical model were challenged by scientific discoveries and new theories. Physics has played an important role here, because its own scientific foundations were altered. Since many other sciences in their admiration for physics followed the classical model, this altered view had a profound effect on them. The main steps in the development of this new paradigm are:

1 *Relativity theory*. The work of Einstein (1905, 1916) was crucial in the process of altering of the classical, mechanical model. He showed that natural forces were variable and dependent on the motion of the observer and that nature could put limits on certain quantities like the velocity of light. This was a first step to bring the influence of the observer back into science. Natural forces were interrelated and unified. Energy and matter were only different forms of the same thing. The maximum velocity of light implied a serious limitation to our abilities to observe.

2 *Field theories*. The idea of action at a distance has been made acceptable by field theories. The first one was electromagnetism (M. Faraday and J. C. Maxwell) although the full understanding of the field concept came later, mainly through the general theory of relativity (Einstein 1916).

3 *Quantum mechanics*. A further decline of the classical concepts came from quantum mechanics, which proved that many natural phenomena were of a discrete nature. It started with Planck (1900), Einstein (1905) and others. It proved the probabilistic nature of the physical world. The indeterminacy principle added an important element to the re-evaluation of the position of the observer. Quantum mechanics described physical phenomena in an utterly different manner from that arising from our daily life experience. For example that the fundamental units of light can be waves or particles, depending on how you look at them, or that particles can influence each other over vast distances.

4 *The expansion of the universe* (Hubble 1929) (see section 4).

5 *Cybernetics*. Our concepts of cause have been enlarged by cybernetics with the notions of interdependency and feedback. Cybernetics has contributed to a deeper understanding of how whole systems operate. More complex forms of causation and regulation of forces can be studied.

6   *Chaos theory*. Chaos theory stressed the importance of instability. It proved that stages of chaos and stability are parts of the same processes and can be studied in a formalized manner. It appears that there are self-organizing forces in nature which can be seen in physical or chemical reactions such as the formation of crystals and living systems (see the chapter by Goodwin in this book). The idea of the existence of self-organizing forces threatens the notion of simple causality as promoted by the classical model.

7   *The discovery of quarks*. Physicists discovered that nuclei consist of protons and neutrons and that these are composed of smaller particles which are not directly observable. In the search for smaller particles, scientists became aware of the influence of their own theoretical concepts upon their research. At the same time, new developments seem to hint at holism.

Many scientists agree that a new paradigm is emerging. It is more difficult to say what it is or will be when it is concrete and mature, because it is still evolving. The proponents of this new paradigm express a sweeping critique of the narrow vision of the classical model, and many hope to repair its negative consequences. For the physical sciences, it is important that the new paradigm facilitates the study of more complex processes, stimulates the development of new forms of mathematics and theories (catastrophe theory, fractals, etc.) and leads to a better understanding of many physical phenomena (waves, turbulences, fields, laser waves, etc.).

In the second part of the twentieth century, we have begun another leap into the unknown. Genetic engineering enables us to cross the boundaries of human life and astronauts took the first steps to investigate the solar system *in situ*. At the same time, in contemporary science we work in regions beyond that of our senses, and thus of all classical and current scientific views. Only the velocity of light and Planck's constant put limits to the observation of the world. Obviously, the distance between our conceptions of the world as built through our daily contact with our visible environment and our scientific reconstruction of it, is growing further all the time. We are truly in a stage of transition from an old towards a new paradigm, a chaotic stage. Attempts by scientists to draw a picture of the new paradigm can be considered a part of the self-organizing tendencies in the scientific community to adapt to new circumstances.

Summarizing, one may say that many scientists feel that science is moving away from the strict determinism, reductionism, positivism and behaviourism of a half-century ago towards new forms. It is still evolving. We see this clearly in some areas of physics, such as quantum mechanics, chaos theory or macrophysics. We have to be cautious in assessing the real impact of these new ideas on the natural sciences so far. For the consequences of this new paradigm may be ambiguous. On the one hand, we see the development of theoretical constructs which allow much more indeterminacy and much more tolerance for divergent viewpoints. On the other hand, DNA research or computers allow much more control of mankind, far from the mind–body discussion in medicine and the human sciences, or the developments in human consciousness. We will have to live with contradictions between these new ideas and make attempts to reconcile them. Presumably science will remain a very open system whose evolution no one can predict. It will continually face the unknown and challenge all old concepts. To support such an evolution, we shall require our intuitive powers to be more closely in accord with our logical and intellectual ones. We need a more synthetic attitude.

3 LESSONS FROM PHYSICS FOR THE FOUNDATION OF A NEW SCIENCE OF HEALTH

Physics as it is conventionally regarded today can provide several notions relevant to a renewal of concepts in the health sciences. Two examples follow.

3.1 Compromising reductionism

Physics is generally accepted as the basic natural science. The classical tradition was to a great extent based upon reductionism. Nevertheless, the proponents of reductionism cannot deny that reductionism can only be applied to a limited range of problems. This can be seen in the relation between physics and chemistry. It is accepted as a fair statement that physics provides the basic explanations of chemistry. Few scientists will doubt this. And yet, what does it mean? Physics, essentially quantum mechanics and electromagnetism, has succeeded in explaining with great precision only a few atoms or molecules of the simplest kind, such as the hydrogenlike atoms (hydrogen, deuterium, tritium). In addition some general rules and properties have been deduced. However, there are millions and millions of chemical molecules, totally unexplained by basic physics alone. Does this imply that one should doubt the viewpoint that physics is the basis for all chemical molecules and reactions? Not at all: if ever someone finds a chemical fact which clearly contradicts basic physics, it would probably mean some new physical law in the making.

However, although fully willing to believe in principle in the reduction of chemistry to physics, it is clear that from any point of view except the most basic one chemistry is another science, with its own laws, its own entities, its own world. This implies that one has to be very cautious about the unlimited ambitions of some proponents of reductionism.

Hence even if we do accept reductionism as a valid scientific view, with due hindsight, it needs modification. Let me call this viewpoint *comprom-*

ising reductionism. Compromising reductionism accepts the validity of reductionist scientific research to investigate certain problems, but adds to that the awareness that reductionism is only a scientific strategy, and not an ontological assumption. In particular we agree with the new model, that all levels are interconnected and that the reductionist approach can only be applied to levels which can legitimately be conceived as seperate layers of reality.[2] Hence one may speak of a *compromising and interconnected reductionism* as a valid strategy for research. In the health sciences too, such a compromising and interconnected reductionism is needed to generate good empirical research to solve many current health problems. If one accepts an interconnected reductionism, interdisciplinary co-operation is unavoidable as soon as two or more levels are studied.

3.2 Physics as the prototype of an axiomatized natural science

The various kinds of mathematics, each with its own consistent system of axioms, constitute perfect examples of axiomatization, i.e. the use of a formal logical system to structure scientific knowledge. The power of an axiomatic system is based on the fact that from a few axioms one may deduce a great number of mutually consistent theorems. Thus unification and generality, together with a kind of transparency (due to the small number of axioms and to the hierarchy of axioms and theorems) are among the most prominent advantages of axiomatized sciences. Axiomatization started with Euclid some 300 years BC. Physics followed nearly 2,000 years later with a first axiomatization by Newton (1687). Axiomatization here means a mathematical modelling of reality or some aspects of it, without necessarily having anything to do with reality. Other natural sciences (chemistry, geology, biology) followed comparatively soon, however, with a much lesser degree of axiomatization. Present-day physics is quite well axiomatized, although it still uses different sets of axioms according to the particular branch of physics which is considered: e.g. gravitation in general relativity, quantum mechanics, thermodynamics, electromagnetism, etc. Some unifications have already been achieved. However the aim is to obtain a unified theory with only one set of axioms from which all branches of physics can be deduced by appropriate approximations (Einstein 1960, Misner *et al.* 1973).

Physics continues to stand up as the prototype of an axiomatized natural science. Thus it is clear that physics in particular, but other natural sciences like chemistry as well, may be of help in constructing a formal basis for a science of health. However, it is possible that one needs first some relationships and rules (and some verification of them) before one really can start to put a science of health into a more or less canonized form – provided this is at all possible. An axiomatization, even at a limited level, may still be as far away in such a health science as it is in medicine itself. Hence, although physics is very illustrative and very fundamental, its immediate usefulness for the foundation of a science of health is uncertain. It would be valuable to try to use axiomatization as a method to structure the basic premises of the health sciences[3] in a consistent way.

4 POSSIBILITIES FOR THE NATURAL SCIENCES TO SUPPORT THE DEVELOPMENT OF A NEW SCIENCE OF HEALTH

4.1 Physics, the industrial revolutions, ecology and health

Needless to say, physics, and the sciences in general, have changed our societies tremendously. Newton's *Philosophiae Naturalis Principia Mathematica* (1687) has changed the face of the world much more than any war. The industrial revolutions were to a large extent direct or indirect consequences of it. And anything that has to do with health has thus been affected. The effect is the greater since physics influenced the structure and evolution of all other sciences (although not always in the best way!).

Ecology is and will be of extreme importance for health, in particular on a worldwide scale. A few examples of the possible contribution of physics may be mentioned here. It will be clear that the solutions will be multidisciplinary in nature: chemistry in particular and other natural sciences will need to join with physics in their study.

(a) The problem of the ozone layer.
(b) The poisoning by heavy metals, and pollution of groundwater.
(c) The greenhouse effect.
(d) Problems with radioactive materials.
(e) Nuclear fusion which will become of paramount importance.

Since the possibilities of nuclear physics are not so well known, some more information follows. Nuclear fusion means the joining of (very) light hydrogenlike nuclei (hydrogen, deuterium, and/or tritium) to form heavier nuclei, in particular helium. A large amount of energy is released in this way. Fusion is the energy source of the H-bomb and of the stars (remember that coal energy is due to sunlight!). Fusion is the opposite of nuclear fission where very heavy nuclei (e.g. uranium) are split and yield energy. Fission is the energy source of the A-bomb and of the present nuclear reactors.

There has been an interest in fusion as an energy source for about four decades and it is being carried out on a scientific basis in several laboratories. However, it will still require at least two decades before its industralization starts to have impact. Tremendous amounts of money are spent to achieve fusion. Only the biggest nations can pay for it: the USA, the European Community, the USSR, Japan and to some extent India, although many countries have smaller programmes of great value. Nuclear fusion seems extremely promising: it will yield cheaper energy, the source is inexhaustible (for millions of years), the nuclear waste will be 20, 30 or 50 times smaller than with nuclear fission (although still a problem!), and it can never become a nuclear bomb. Its effect could be beneficial too for the Third World countries because of the cheapness of the energy.

Nuclear fusion is very promising in all respects and will have a major impact on the world as a cheap and reasonably ecological energy source in the decades after 2020. It is an ecological topic of first rank and could be a significant contribution of physics and engineers in developing a more healthy physical and chemical environment. It is clear that a science of health may require that physicists work in programmes which have ecological relevance. Physics can also contribute to the development of alternative forms of energy, such as solar or wind energy, which will bear fruit mostly on a short or intermediate time scale.

4.2 Methods of physics which may be useful for a science of health

Physics has, in the past, been very influential in health sciences in the development of disciplined research methods. Some new methods, emerging as a part of the development of the new paradigm in the natural sciences, could stimulate innovative research in the health sciences. In particular, developments in mathematics have made it possible to treat discrete and non-linear relationships in an exact, formalized way. This new mathematics[4] is at the basis of chaos theory, fractals, etc. The new mathematics and its applications make it possible to investigate patterns of high complexity, and as such to come closer to reality. This could be important for the health and human sciences because it offers a certain alternative for overly simple causal relationships (as in the Stimulus–Response model). Such methods are already in use in biology and (neuro)physiology. For example:

- It has been shown that the seasonal variations of the measles epidemics exhibit underlying chaotic behaviour. The underlying order is describable by the mathematical formulae of chaos theory (Schaffer and Kot 1985).
- The fluctuations of the human heart rate have characteristics which are attributable to underlying chaotic behaviour (Goldberger 1987; Janssens 1990).
- The normal central nervous system shows evidence of chaos (Babloyantz 1988).
- Chaos theory and fractals are used in working with some measuring instruments (e.g. electro-encephalography).

Such new methods of analysis have helped our understanding of diseases such as depression, dementia, epilepsy, diabetes, and the cognitive abilities of the brain. In particular, all phenomena in which very complex patterns of rhythm occur can benefit from these new mathematical methods.

It is also shown that certain chaotic irregularities (as in the case of the heart rate or the nervous system) are very normal, and that regularity can be a signal of a serious disease (e.g. in the case of ventricular fibrillation, depression or epilepsy). This is explained by the fact that chaotic states seems to be information-rich whereas simple, regular states are monotonous, with loss of flexibility (Kyriazis 1990). To support further developments in this field, a close co-operation between mathematicians, (bio)physicists and clinicians is urged.

4.3 Further support from physics

In the past, physics has had a profound influence on medicine, especially through innovative applications. There are many examples of this in the field of high-tech medicine: X-rays and all kinds of medical applications of radioactivity, thermography, laser-technology in surgery, echography, computer hardware and all its applications, CAT and PET scanners based upon high-energy physics, etc.

For the future one may expect that many innovations in physics will have important spin-offs in the health sciences. We no longer make the classical error of thinking that we already know all natural forces, energies or effects in physics. We can be quite sure that new forms of physical phenomena will be discovered as we penetrate more and more into the universe and the microcosm. These may, as in the past, drastically reshape our conceptions about health and disease. If new forms of energy were to be discovered (see Goodwin, in this book) – as was the case with radioactivity in the past – it would have a major impact. Some topics in this field are still highly speculative (such as psi energy), but it is a real point of discussion and challenge to the scientific world to prove or reject the existing intuitions about new energies or known energies with unexpected effects.

The philosophy of the natural sciences could be helpful to the development of new paradigms by studying processes of creativity. The classical model suggests an over-logical concept of scientific activity. Science was conceived as the result of a kind of logical algorithm. As is commonly known, intuition, creativity, and even spirituality played an important role in the work of many outstanding physicists like Newton, Einstein, Bohr, Pauli, Bohm and others. The influence of these qualities on their work may be underestimated. The axiomatized language of physics has buried these aspects of their work. It could be a target of the philosophy of science to highlight these qualities of creativity and search for a way to understand them.

As we penetrate more deeply into the macro- and microcosm, much more awareness is growing in physics about the influence of researchers on the subject of their research. This is an important point, especially for the health sciences (see many of the chapters in this book) and all human sciences. More research on this relationship will have a double function. First, it can help to legitimate research on this topic in other sciences. Secondly, physics can be important in investigating the limits of the influence of the researcher, because in other scientific fields, the subjects of research react (relatively) autonomously. In physics, the situation might be more easily investigated.

5   CONTRIBUTIONS TO OUR UNDERSTANDING OF WORLDVIEWS

Health conceptions are related to existing worldviews. Therefore, the topic of worldviews is a central issue in the development of a new science of health (see the chapters by Lafaille and Göpel in this book). Physics has contributed to the changes in our worldviews and to the development of a more coherent and empirical view of the world. In this section we will discuss some elements which might be important for the development of a science of health.

5.1 Changing cosmological worldviews

The worldview contains the ideas which a person has concerning the connection between the physical world, particularly at its largest and its microscale, and a more spiritual world, in particular God (including the denial of God), the community, etc. Obviously the worldview has to deal with natural sciences (which study reality), with religion, with philosophy and even with social sciences. It is full of unanswered questions, some of them not well formulated, some belonging to physics and others to metaphysics. For example: Is the universe finite in space and time? Did it have a sudden origin? Why is nature as it is? Why is a mathematical description of nature possible? Why are there three space dimensions and one dimension for time – or is this only apparently so? How do these dimensions influence the mind or health? How could the universe be created and, if God was the creator, how did he come into existence and are there laws to whom he is submitted? What is the purpose of mankind?

The eighteenth and nineteenth century and even part of the twentieth century lived in the enchantment of the great Newtonian idea. Newton axiomatized physics and thus united many phenomena into one theory. For example, earthly and celestial physics were united. He cast a mathematical framework to describe causal relations in a fully deterministic way. The impact of being able to perfectly predict the behaviour of any mechanical system was very great. Philosophy (which still included physics and the other natural sciences at Newton's time) left physics and the other natural sciences and turned to mankind. If the whole world was a perfect clockwork, what was the human being doing in this clock? What was left over of free will in this fully deterministic system? In the nineteenth century the idea of evolution in biology (Darwin) started to alter the prevailing Newtonian worldview (see above). The twentieth century brought great shocks too: the special and the general theory of Einstein connecting space and time and matter, the idea of an unbounded but finite universe, the Big Bang, quantum mechanics (see above), and so on. The present-day worldview is in full evolution. If the worldview is changing drastically, then most people feel disoriented: a vague world picture is sometimes more comforting than one which is unstable and quickly overthrown, especially as this confronts us again with so many unanswered questions. The moral and ethical depressions of the present time are partly related to the instability and variability of the worldviews. And yet this is part of the price to be paid for a new insight.

5.2 Evolution and the arrow of time

The idea of time and evolution is generated by life itself, by experience, by history, by biology, by geology, by astronomy and cosmology. We feel that time evolves, as we can distinguish past from present and future; we feel 'the arrow of time'. Often we would like to stop time. In particular, in view of our ultimate fate, we would like to stop or at least to delay the rush of time as much as possible and in this respect health is central.

There is a paradoxical situation connected with the concept of time. On the one hand all the basic equations of physics allow time inversion, i.e. whether time flows in one direction (the future) or in the opposite one (the past), the equations do not change. And yet we do experience the arrow of time, we do distinguish past and future. The explanation of the paradox is based on statistical physics. Yet, it hints at the possibility of prolongation.

The idea of determining the origin of time goes back to Kepler, who used the Bible to fix the creation around 6,000 years ago. Others (Archbishop Usher, Newton) indicated similar time scales until geology put an end to speculations based on the Holy Scriptures. James Hutton and Charles Lyell may be mentioned here. We are now fairly confident (using radioactive decay of ores, etc.) that the Earth came into being some 4.57 billion years ago and that the universe has existed for at least some 15 billion years.

Copernicus (his book was published in 1543) taught us that the Earth was not the centre of the universe (which at that time was conceived as limited to the solar system only), but that the Sun was. The change from a geocentric system to a heliocentric one is known as the Copernican revolution and required an adaptation of the human mind: it meant something to have to accept that your home is not anchored at a fixed place. This drastically influenced the worldview of the following period. In the beginning of this century Shapley showed that the Sun was not at all in the centre of our galaxy (the Milky Way) and some decades ago Baade and others made it clear that our galaxy, although a big one, is not at all among the biggest (some have diameters that are 50 times larger). Thus our galaxy by no means represents the centre of the universe.

Moreover, Hubble in the years 1920–30 and later, showed that the universe is expanding. Hence our home at its very biggest scale is not at rest but in continuous motion. It requires some psychological suppleness, to realize that one lives in an ever-changing world, even if the space and time scales are very much larger than the lifespan of a human. Moreover, we do not know whether the universe is ever-expanding or whether it is oscillatory, i.e. after a maximal expansion it may shrink again, and whether this happened several times in the past and may happen several times in the future. In the hypotheses of the Big Bang the universe started as an infinitely small singularity and exploded. Although this theory is very valuable to describe a large part of cosmological history, I can not accept the limit of a singularity. The smallest volume may have been very small (relatively speaking, maybe it was still as big as a galaxy, i.e. about 100,000 light-years in diameter!) but not a point containing all mass in a singularity as understood in the conventional sense. I have forwarded an alternative theory, i.e. the Quiet Big Bang, based on the cosmological zero energy–momentum principle. This is the hypothesis that the total positive and negative energy is zero at the beginning of the Universe and at all times. Both the positive and negative energy grow simultaneously with the expansion of the universe, their sum remaining perfectly zero. This avoids postulating the singularity with an infinite density, yet retains the advantages at the later stages of the ordinary Big Bang theory. Several arguments favour this new viewpoint. One of its benefits is that the huge amount of (positive) energy or matter presently in the universe need not be assumed to have been created at once and without counterpart. Moreover, it is much more plausible that a quantum fluctuation created the Quiet Big Bang rather than a huge amount of positive energy. The hypothesis makes the gigantic task of the Creator (whether this is God or a metaphysical principle) much easier, although by no means light. Indeed, although starting from zero, it was a great event. Capra (1975 and 1982) has pointed out that these considerations are very much in accord with Buddhist views on the inherent emptiness of existence, and they may now influence the Western mind accordingly.

5.3 Quantum mechanics and determinism

According to quantum mechanics one cannot simultaneously and precisely determine the place of something and its momentum (its mass times the velocity). There is a kind of indeterminism. You cannot know everything perfectly: the more precisely one property is known, the less precisely the other (the conjugate property) is known. The product of both 'errors' has a lower limit determined by Planck's constant. This constant is very, very small and the effects, while dominant for the microcosm, are nearly always irrelevant on a macroscopic scale. This lower boundary is uncomfortable and surprising, but quantum mechanics explains thousands of experiments, so that one can no longer doubt it: if not perfectly correct then quantum mechanics is at least an extremely good approximation with no known flaws.

Many questions, some metaphysical arise here: is this indeterminacy due to an inadequacy of measuring (one measurement perturbs the system, hence the second measurement is affected) or is this indeterminacy inherent in nature itself? It may be mentioned here that Einstein, who contributed strongly to quantum mechanics himself, never fully accepted it. He made the famous statement 'God doesn't play dice'. He could not accept the inherent indeterministic character of quantum mechanics. Einstein agreed that quantum mechanics is a self-consistent mathematical system and that it is confirmed by many, many experiments. However, just as he had not been able to agree with the logical foundations of the gravitational theory of Newton (although it was a very successful theory indeed) replacing it with his own gravitational theory, so he wanted an improved quantum mechanics.

Quantum mechanics has a destabilizing effect on our current worldviews and hence indirectly on a science of health. Moreover, chemistry is based on quantum mechanical effects and chemistry is all-important to our cells and to our health.

The quantum mechanical indeterminacy has revived the question of free will. Yet the total determinism of the Newtonian way of looking at life was already weakened by probabilistic arguments.[5] Some quantum phenomena show extraordinary forms of movement of particles. They do not move through space in a continuous manner, but in steps as if they appear, disappear and reappear. This is seen in the notorious photon experiment in which it is impossible to decide how a photon passes through two gates in a screen (see Lijnse 1981: 51–66). Such phenomena urge us to rethink some of our daily concepts about reality. In particular, our notions of determinism have to be enlarged and are challenged by quantum mechanics. Such discoveries may have a great impact upon our current thinking and upon questions like the consequence of illnesses, the existence of free will, and so on, which are so crucial for the health sciences. On the other hand, there is a vast literature of speculations on these topics, and it is clearly still too early to draw definite conclusions.

6 CONCLUSION

We have described the evolution in physics from a classical to a new model. This evolution forms a background for some considerations about the foundations of a new science of health. The first topic dealt with reductionism: the ideas of compromising and interconnected reductionism were proposed. In this way, the advocates and the opponents of reductionism may be brought closer together. The second topic dealt with physics as the prototype of an axiomatized natural science. One should try to clarify the basic principles of a science of health as soon as possible, although, even with the example of physics in mind the task is not easy. Further, some possibilities for physics to improve health were discussed. The importance of physics for the solution of ecological problems is clear. However, the different fields are rather autonomous at present. Nevertheless, it seems important to study the connection: a science of health should have enough links with physics to influence research, industry and ecology. A next step was to consider which methods of physics could be useful in establishing a science of health.

The last paragraph concerns the influence of physics on our worldview. As discussed in earlier chapters, the quesion of worldviews is crucial in the development of a science of health. Such discussion has to be multidisciplinary of nature and we looked at the possible contribution of physics. We have discussed this topic in three sections: the first one deals with cosmology; the second with evolution (the Quiet Big Bang, biology (Darwin); the arrow of time); the third deals with indeterminism in quantum mechanics as well as in classical physics as a consequence of statistical effects. As a final conclusion we may state that there is sufficient need to establish a good link between physics and a science of health. However, as far as immediate results are concerned, hopes should not be too high. Nevertheless a start should be made.

REFERENCES

Arnodl, Vl. (1983) *Catastrophe Theory*, Berlin/Heidelberg: Springer-Verlag.
Babloyantz, A. (1988) 'Chaotic dynamics in brain activity', in E. Bazar (ed.) *Dynamics of Sensory and Cognitive Processes by the Brain*, Berlin: Springer-Verlag, pp. 196–202.
Capra, F. (1975) *The Tao of Physics*, Berkeley: Shambhala.
—— (1982) *The Turning Point: Science, Society, and the Rising Culture*, New York: Bantam Books.
Close, F. E., Marten, M. and Sutton, C. (1987) *The Particle Explosion*, Oxford: Oxford University Press.
Davies, P. (1990) *The New Physics*, Cambridge: Cambridge University Press.
Einstein, A. (1960) *The Meaning of Relativity*, London: Methuen & Co.
Einstein, A., and Infeld, L. (1950) *Die Evolution der Physik*, Wien/Hamburg: Paul Zsolnay Verlag.
Glass, L. and Mackey, M. (1988) *From Clocks to Chaos – The Rhythms of Life*, New York: Princeton University Press.
Gleick, J. (1988) *Chaos*, London: Heinemann.
Goldbergher, A. (1987) 'Non-linear dynamics, fractals, cardiac physiology and sudden death', in L. Rensing and Van Der Heiden (eds), *Temporal Disorder in Human Oscillatory Systems*, Berlin: Springer-Verlag.
Harman, W. W. (1992) *A Re-examination of the Metaphysical Foundations of Modern Science*, Sausalito, CA: Institute of Noetic Sciences.

Janssens, L. (1990) *The Integral-Differential Equations on the Ventricular Cardiac Suction Pump. A Catastrophe Theoretic Approach to the Matching Stimulation Method*, diss., Antwerp: University of Antwerp.

Kyriazis, M. (1990) 'Dynamical systems in medicine', *Scientific and Medical Network, Newsletter* 44.

Lauwerier, H. A. (1987) *Fractals*, Amsterdam: Aramith Uitgevers.

Lijnse, P. L. (1981) *Kwantum-Mechanica. Een eenvoudige inleiding*, Utrecht/Antwerp: Het Spectrum.

Longair, M. S. (1984) *Theoretical Concepts in Physics*, Cambridge: Cambridge University Press.

Mandelbrot, B. (1982) *The Fractal Geometry of Nature*, New York: W. H. Freeman.

Misner, C. W., Thorne, K. S. and Wheeler, J. A. (1973) *Gravitation*, San Francisco: W. H. Freeman & Co.

Peitgen, H.-O. and Richter, P. H. (1986) *The Beauty of Fractals*, Berlin: Springer-Verlag.

Rapp, P. E., Bachore, T. R. and Zimmerman, I. D. (1990) 'Dynamical characterisation of brain electrical activity', in Krasner, S. (ed.) *Ubiquity of Chaos*, New York: AAAS.

Saunders, P. T. (1980) *Introduction to Catastrophe Theory*, Cambridge: Cambridge University Press.

Schaffer W. M. and Kot, M. (1985) 'Nearly one dimensional dynamics in an epidemic', *Journal of Theoretical Biology* 112: 408–27.

Schilpp, P. A. (1957) *Albert Einstein: Philospher-Scientist*, New York: Tudor.

Schrödinger, E. (1963) *Space-Time Structure*, Cambridge: Cambridge University Press.

Wildiers, M. (1973) *Wereldbeeld en teologie. Van de middeleeuwen tot vandaag*, Antwerp/Amsterdam: Standaard Wetenschappelijke Uitgeverij.

—— (1988) *Kosmologie in de Westerse Cultuur. Historisch-kritisch essay*, Kapellen: DNB/Uitgeverij Pelckmans.

5

HEALTH IN MEDICAL SCIENCE
From determinism towards autonomy
Rudy Rijke

The recognition that many factors other than disease influence health is a step toward understanding the present widespread disappointment with our medical system. Our previous supposition that health is necessarily the result of technologically accurate interventions into human physiology and biochemistry or the outcome of a precise therapeutic response to pathology has not proved to be congruent with our experience. The common expectation that the health care system through the use of knowledge and skill should be able to 'produce' health has not been fulfilled. To some extent, it is not only the nature of the health care system but the nature of our expectations and the assumptions on which they are based that require a thoughtful re-examination.

(Remen 1980)

1 'HEALTH' IN MEDICAL SCIENCE

The knowledge of present-day medicine has been obtained by studying diseases and the consequent effect of therapeutic methods upon them. The principal aim of medical research is to determine which factor (or factors) causes a particular disease. It may be a chromosomal aberration, a bacterium or a virus, or a certain life-style or life conditions. Ideally, the mechanism by which the causal factor affects the body is mapped so that methods can be developed by which the influence of the causal factor is prevented or lessened, or its effect on the body reduced.

The implicit assumption underlying this approach is that health is the absence of deviations and complaints and that biochemical processes in the physical body primarily determine the state of health. Another assumption is that the occurrence of a disease is determined by a causal factor and that the elimination of the disease is achieved by a certain therapy if one is available. Ideally, all 'causal factors' of diseases, or, put more generally, all 'unhealthy factors' would be eradicated from the world. This has been done successfully with smallpox and the assumption is that something similar can be done with all diseases. Yet another assumption is that people are identical in these respects, so these 'causal factors' and therapeutic methods have the same effect on all.

However, medical science itself has shown that health is not dependent on the absence of 'unhealthy factors'. We are subjected daily to a host of bacteria and viruses, to chemicals and irradiation. We all continuously suffer micro-traumata with each movement, even in sitting or lying down. We are all subjected to various forms of psychological trauma and to the 'small hassles of life'. So, considering these and many other 'unhealthy factors' and traumata, it is no wonder that Peck speaks of the 'miracle of health' (Peck 1978). It also becomes clear that health includes natural healing processes on all levels of human existence (de Vries 1985).

2 'HEALTH' IN THE GENERAL POPULATION

Very little is known in medical science about the natural history of diseases, complaints and healing processes in the general public. Most medical research is being done in academic institutions and hospitals where a select group of the population with certain abnormalities, complaints and diseases is seen.

An example of this is the difference found by Ellenberg and Nelson (1980) in the frequency of unfavourable sequelae in a common disorder of childhood, febrile seizures, between 7 population- and 19 clinic-based studies. In population-based studies it was found that 1–4.5 per cent of children experienced later non-febrile seizures (epilepsy) after one or more febrile seizures in childhood, whereas this may be as high as 80 per cent in clinic-based studies.

Large differences even exist between the general population and the people who consult a general practitioner, as has been found by Huygen and co-workers (1983). By thoroughly examining a large sample of a general practice, including all age groups, they found that although approximately 90 per cent of the people considered their health to be good or excellent, 50 per cent had 4–10 physical complaints and 20 per cent had more than 10 complaints (5 per cent having no complaints). These complaints were not just minor nuisances. It appeared that only 10 per cent of the people with stomach pains consulted the general practitioner, only 10 per cent of the people with chest pain, 10 per cent with a productive cough, 15 per cent with lumps and 25 per cent with abnormal blood loss.

In another study, X-rays were taken of the vertebral column in people with low back pain and in a matched group from the general population who never had low back pain (van der Does 1980). It was found that 73 per cent of the men and 81 per cent of the women with low back pain had deformities in the vertebral column. Of the people without low back pain this was 66 per cent and 81 per cent, respectively.

So, from research done in the general population, it appears that health may exist in spite of physical complaints and even physical abnormalities.

A rather extreme example of this was shown by Lorber (1980) who, using computer tomography, examined the brains of people on whom he had operated in their early youth because of hydrocephaly. He found that some of these perfectly normal, intelligent young people had less than 5 per cent of the normal amount of brain tissue left.

Also, in a large prospective study in a prenatal care unit including all children (more than 37,000) it was found that neurological damage is not always permanent (Nelson and Ellenberg 1982). More than half (51.5 per cent) of the 229 children who were diagnosed as suffering from cerebral palsy at the age of one year were free of motor handicaps at age 7. It was also found that 25 per cent of the black children and 13 per cent of the white children whose motor signs resolved were mentally retarded at the age of 7 years.

Findings like these show that we know very little about the natural history of physical abnormalities and that these abnormalities will not necessarily impair health, functioning and well-being.

The application of knowledge obtained by a deterministic medical science may also prove hazardous, as has been shown in several intervention trials for coronary heart disease. In a study in 53 clinical centres in the USA all men between 30 and 64 years old with electrocardiographic evidence of a myocardial infarction that had occurred less than 3 months previously were randomly assigned to a group receiving a cholesterol-reducing drug (clofibrate) or to a placebo-group (Coronary Drug Project Research Group: 1980). Good adherents to clofibrate (patients who took more than 80 per cent of the protocol prescription during the 5-year follow-up period) had a substantially lower 5–year mortality than did poor adherers to clofibrate (15.0 versus 24.6 per cent). However, similar data were obtained for the people in the placebo-group: 15.1 per cent mortality for the good adherers and 28.3 per cent for the poor adherers.

Such findings show that reducing the factors that were previously thought to 'determine' the occurrence of cardiovascular diseases does not give any improvement in groups of the general population. So, either other factors are more important in the occurrence of cardiovascular diseases, or therapy and intervention have little value when applied to the general population.

What all these findings seem to imply is that health (and healing) is much less determined by 'unhealthy factors', physical complaints, abnormalities and 'risk factors' than we previously assumed based on the findings of medical science. It also seems to imply that reducing these factors does not necessarily lead to better health, when applied to the general population. Moreover, it is becoming clear that studying the factors that seem to 'determine' the occurrence of a certain disease does not necessarily lead to knowledge of, or the promotion of, health and healing.

3   TOWARDS A SCIENCE OF HEALTH

Antonovsky (1987) recently stated that there is a fundamental difference between a science which studies diseases and a science which studies health: the questions, hypotheses, methods and the answers and conclusions will be different. For instance, research into the mechanisms and factors of how smoking may lead to lung cancer is very different from the research which tries to find out why most smokers do not get lung cancer.

An interesting example of this difference is the comparison of two recent publications on the health of elderly persons. Rozzini and his co-workers (1988), using a disease-oriented approach, investigated the relationship between somatic symptoms, depression and life conditions (health status, function, social satisfaction, income) in a group of 1,201 elderly people living at home. They found depression to be the most important factor in the occurrence of somatic complaints. Life conditions were important co-factors in defining well-being. They looked for the factors leading to somatic complaints and found many factors operating. Their conclusion is the usual result of this kind of research:

> The results suggested that positive life conditions may buffer complaints of somatic origin in non-depressed elderly people. For this reason, in organizing programs for the well being of elderly people it is important *to consider all aspects of life* which may influence somatic health [my italics].

A health-oriented approach was followed by Rohe and Kahn (1987), who stated that research on ageing has been focusing too much on the average, on the usual. Usually, elderly people develop diabetes (caused by a decreased sensitivity of the peripheral tissues to insulin), osteoporosis, a decline in physical and mental functioning, and so on. However, Rowe and Kahn raise the question: how is it possible that some elderly people do not show any signs of diabetes, osteoporosis, or decline in physical and mental functioning? From dozens of publications, it appears that people who do not show the 'usual signs' of ageing, differ in being physically and mentally active.

It also appears, that increasing the level of activity in people with the 'usual signs' of ageing leads to the disappearance or reduction of these signs. When research was focused on how elderly people remain healthy, they found two main factors: 'autonomy' and 'social support' (if the latter was 'autonomy-enhancing'). It was also found that if measures are taken (e.g. in homes for the elderly) which promote autonomy and social support, the number of physical complaints and the frequency of physical illness are reduced.

It is clear that a health-oriented approach not only leads to fundamentally different research findings, but also has very different consequences.

Similarly, investigating healing processes (an integral element of health) leads to different answers and consequences than investigating diseases and their treatment. Some examples of such research with cancer patients follow.

Lerner (1985) investigated all centres for 'complementary cancer therapies' in Europe and North America. He intended to get an overview of the various complementary approaches that are being used by people with cancer; of the kind of people who make use of them, and what the effects are of these complementary approaches. His findings were:

1   There is no effective 'treatment' among the complementary approaches.
2   These approaches are 'health-promoting'.
3   It seems that 85–90 per cent of the people benefit from them.
4   Each individual needs to find and develop his/her approach.

This author (Rijke 1985) found that the course of action taken by 'exceptional cancer patients' (people with a serious form of cancer who live longer and better without effective treatment, and with partial or complete remission of their cancer), was very different and individualistic.

LeShan, who has been investigating healing processes in people with cancer for more than 40 years, also stresses the importance of the autonomy and the individuality of people in promoting health and healing (LeShan 1989). In describing his early experiences in psychotherapy with cancer patients, he says that it was interesting and rewarding (most people were very co-operative), but that three observations started to bother him:

1   What happened in the therapy had little relevance to the daily life of these people.
2   He saw things in the therapy that did not fit into his theories (courage, dignity, and so on).
3   The people did not get better.

He changed from a disease-oriented towards a health-oriented approach:

Every psychotherapy process, early in its development, defines the basic questions it is trying to answer. In the usual models of therapy, there are three of these:

1   What is wrong with this person?
2   How did she or he get that way?
3   What can be done about it?

These questions are central to the therapy process, whether the therapist has been trained in a Freudian, Jungian, Adlerian or humanistic manner. Therapy based on these questions can be wonderful and effective for help with a variety of emotional and cognitive problems. *It is, however, not effective with cancer patients. It simply does not mobilize the person's self-healing abilities and bring them to the aid of the medical program*. We have now had enough experience in many different countries to state this as a fact.

The therapeutic approach developed in this research, for work with people with cancer is based on entirely different questions. These are:

1   What is right with this person? What are his/her special and unique ways of being, relating, creating, that are his own and natural way to live? What is his special music to beat out in life, his unique song to sing so that when he is singing it he is glad to get up in the morning and glad to go to bed at night? What style would give him zest, enthusiasm and involvement?

2   How can we work together to find these ways of being, relating and creating? What had blocked their perception in the past? How can we work together so that the person moves more and more in this direction until he is living such a full and zestful life that he has no more time or energy for psychotherapy?

(LeShan 1989)

So, LeShan and others discovered that it is necessary for the promotion of healing to address the individual uniqueness of people. Of course, individuality poses a great problem for a deterministic medical science which is trying to establish 'general laws'.

4   WHAT IS HEALTH?

The World Health Organization's definition of health is 'a state of complete physical, mental and social well-being'.

So, what do we find when we broaden our scope beyond the medical world, in which health is always (for all!) so closely related to disease?

Flannery (1989) has been studying people who respond to demanding situations without losing their sense of well-being. For example, he compared the coping styles of night-school students with the most and the fewest episodes of illness during a certain period. It is interesting that he uses 'episodes of illness', which is probably a reflection of the state of health, as a parameter, there being a lack of quantitative parameters for well-being. He found that:

- People with fewest episodes of illness tended to maintain reasonable control in their lives (including when problems came up).
- They were personally committed to a goal of some kind (they spent 4–6 hours a week doing something that provided a sense of challenge and enhanced their sense of meaningful participation in life).
- They used a minimum of drug-like substances (including nicotine and caffeine) and spent at least 15 minutes a day on some form of active relaxation.

- About 80 per cent of them engaged in regular aerobic exercises.
- They tended to seek out other people to be actively and empathically engaged with them.

Sheehy (1981) investigated well-being using a 'well-being test' with more than 60,000 people. From these data she selected a 'high-scoring' and a 'low-scoring' group that she interviewed in depth and followed for some time. She found, among other things, that the 'high-scoring' group usually had the first traumatic experience at an earlier age in their youth than others. She found that these people differed from others in the following characteristics:

courage; faith; creativity; flexibility; ability to love; being a model for people in a crisis; having true friends; conviction that life has meaning; humour; energy.

5 CHARACTERISTICS OF HEALTH

To make research into health and health promotion more concrete, we suggested earlier the following characteristics of health, which are based on findings from research on health and healing (Antonovsky 1987; Bergsma 1984; Cousins 1979; Flannery 1989; Freudenberger and Richelson 1980; Lerner and Remen 1985; LeShan 1989; Lynch 1977, 1985; Remen 1988, 1989; Rijke 1985; Rijke and Rijke-de Vries 1988; Rogers 1980; Rohe and Kahn 1987; Sheehy 1981):

1 Autonomy.
2 Will to live.
3 Experience of meaning and purpose in life.
4 High quality of relationships.
5 Creative expression of meaning.
6 Body awareness.
7 Consciousness of inner development.
8 Individuality: the experience of being a unique part of a greater whole.
9 Vitality, energy.

6 HEALTH PROMOTION AND SCIENCE

From the (limited) data that are available from research on health and health promotion (including the study of healing), it seems that the discovery and development of autonomy is crucial. This process, which has been described elsewhere (Rijke 1985), usually involves an existential crisis out of which a sense of autonomy emerges (Remen 1988; Rijke and Rijke-de Vries 1988; de Vries 1986). Autonomy is controversial and not naturally available to people: the crisis which may lead to the discovery and the development of autonomy, involves a letting go of many of the previously implicit assumptions about health, disease, life and oneself, which used to be the foundation of one's way of life (Rijke 1985).

The thesis of this contribution is that a science of health and of health promotion, including the study of healing, must take autonomy and individuality of people as a primary concern in its studies.

Or, to put it differently, it seems that health and health promotion consist of moving from being determined by 'unhealthy factors' toward autonomy and individuality.

Is this a defeat for medical science? Prigogine and Stengers (1984), discussing the failure of deterministic approaches in the fields of chemistry and physics, wrote:

> Is this a defeat for the human mind? This is a difficult question. As scientists, we have no choice; we cannot describe for you the world as we would like to see it, but only as we are able to see it through the combined impact of experimental results and new theoretical concepts. Also, we believe that this new situation reflects the situation we seem to find in our own mental activity. Classical psychology centered around conscious, transparent activity; modern psychology attaches much weight to the opaque functioning of the unconscious. Perhaps this is an image of the basic features of human existence. Remember Oedipus, the lucidity of his mind in front of the sphinx and its opacity and darkness when confronted with his own origins. Perhaps the coming together of our insights about the world around us and the world inside us is a satisfying feature of the recent evolution in science that we have tried to describe.

According to them, science is in a development in which the unpredictableness of singular events in open systems, the dynamic complexity of reality in which open systems strive toward higher levels of complexity and meaning, are *central*, and in this, science must start to engage itself in a dialogue with nature rather than posing its questions from preconceived theories as a monologue. For a science of health this would mean the unique individuality of people, the dynamic complexity of daily human life, the will to live, health, healing and development, and the necessity for a human dialogue, both within oneself and with one another.

Stuart Miller, one of the originators of the Institute for the Study of Humanistic Medicine in San Francisco in 1971, said about this:

> As in medicine, so elsewhere: the advantage of materialistic science and technology has failed to bring all the blessings it promised. Therefore, our task is now to keep the genuine advances of science and technology and to combine them with a health giving way of life.

In medical terms this means that groups of investigators must begin to concern themselves with the person: not only with the patient, but with his underlying humanness. Not just with his or her disease, but with his body as a whole, and with his feelings, the mind, and the spirit . . . even the spirit . . . of the patient.

(Miller 1974)

REFERENCES

Antonovsky, A. (1987) 'The salutogenic perspective – towards a new view of health and disease', *Advances* 4(3): 47–55.

Bergsma, J. (1984) 'Towards a concept of shared autonomy', *Journal of Theoretical Medicine* 5: 325–31.

Coronary Drug Project Research Group (1980) 'Influence of adherence to treatment and response of cholesterol in the coronary drug project', *New England Journal of Medicine* 303: 1038–41.

Cousins, N. (1979) *The Anatomy of an Illness as Perceived by the Patient*, New York: Norton & Co.

Does, E. van der (1980) 'Huisarts en epidemioloog', *Medisch Contact* 35: 485–90.

Ellenberg, J. H. and Nelson, K. B. (1980) 'Sample selection and the natural history of disease', *Journal of the American Medical Association* 243: 1337–40.

Flannery, R. B. (1989) 'The stress-resistent person', *Harvard Medical School Health Letter* 14(4): 5–7.

Freudenberger, H. J. and Richelson, J. (1980) *Burnout*, New York: Bantam Books.

Huygen, F. J. A., Hoogen, H. van den and Neefs, W. J. (1983) 'Gezondheid en ziekte – een onderzoek van gezinnen', *Nederlands Tijdschrift voor Geneeskunde* 127: 1612–19.

Lerner, M. (1985) 'A report on complementary cancer therapies', *Advances* 2(1): 31–43.

Lerner, M. and Remen, R. N. (1985) 'Varieties of integral cancer therapies', *Advances* 2(3): 14–33.

LeShan, L. (1989) *Cancer as a Turning Point*, New York: Dutton.

Lorber, J. (1980) 'Is your brain really necessary?', *Science* 210: 1232–4.

Lynch, J. J. (1977) *The Broken Heart*, New York: Basic Books.

—— (1985) *The Language of the Heart – the Body's Response to Human Dialogue*, New York: Basic Books.

Miller, S. (1974) 'A new humanism in medicine', *Synthesis* 1:137–45.

MRFIT Research Group (1982) 'Risk factor changes and mortality results', *Journal of the American Medical Association* 248: 1465–77.

Nelson, K. B., and Ellenberg, J. H. (1982) 'Children who "outgrew" cerebral palsy', *Pediatrics* 69: 529–36.

Peck, M. S. (1978) *The Road Less Travelled*, New York: Simon & Schuster.

Prigogine, I., and Stengers, I. (1984) *Order out of Chaos – Man's New Dialogue with Nature*, New York: Bantam Books.

Remen, R. N. (1980) *The Human Patient*, New York: Doubleday.

— — (1988) 'Spirit: resource for healing', *Noetic Sciences Review* 8: 4–9.

— — (1989) 'Feeling well: a clinician's casebook', *Advances* 6(2): 43–50.

Rijke, R. P. C. (1985) 'Cancer and the development of will', *Journal of Theoretical Medicine* 6: 133–42.

Rijke, R. P. C., and Rijke-de Vries, J. (1988) 'Zingeving en verantwoordelijkheid – de subjectieve ervaring van mensen met kanker', *Metamedica* 67: 124–31.

Rogers, C. R. (1980) *A Way of Being*, Boston: Houghton Mifflin Co.

Rohe, J. W. and Kahn, R. L. (1987) 'Human aging: usual and successful', *Science* 237: 143–9.

Rozzini, R., Bianchetti, A., Carabellese, C., Inzoli, M., and Trabucchi, M. (1988) 'Depression, life events and somatic symptoms', *The Gerontologist* 28: 1–4.
Sheehy, G. (1981) *Pathfinders*, New York: Bantam Books.
Vries, M. J. de (1985) *Het behoud van leven*, Utrecht: Bohn, Scheltema & Holkema (Holland).
—— (1986) 'Crisis en transformatie: lessen van wonderbaarlijke patiënten', *Medisch Contact* 41: 715–56.

6

HEALTH AT SOCIAL CROSSROADS

Theories of human interaction and a science of health

Peter Mielants and Paul Rijnders

1 INTRODUCTION[1]

In this chapter we will examine how the perspective of system theory can contribute to the development of the health sciences. We will suggest how to deepen our insight into the way people relate to each other, using system theory. The WHO Ottawa Charter explicitly states that social modes of relation have an important impact on health. Such an improved insight into how people affect each other could therefore be an important new direction in prevention, especially of psychosomatic disorders and psychological problems. In this chapter we suggest that there is one aspect of health in which an individual looks upon his/her personal reality as a product of interactional processes. As opposed to the common view which defines health and illness as characteristics of the individual, it is argued here that symptoms and definitions of health and illness are subsidiary to interactional processes (Minuchin 1973; Wynne *et al.* 1987). By increasing the knowledge and awareness of those human interactions, system theory can help to relate personal views (about oneself, one's partner, the relation, . . .) to the relational network.

The general principles of the system theory will not be elaborated on here, since they are already dealt with elsewhere in this book (see the contribution by Lafaille). What we will do here, is to concentrate mostly on human relationships as they mainly arise in primary groups (family, living groups, small working situations). This will be dealt with in Section 2.

The paradigm of system theory differs from other views on health by two basic assumptions:

1   Health and illness are seen in system theory as a structuring of reality by descriptions which are continuously (re)created in human interactions. System theory thus postulates a far more process-like idea of health and

illness. The connection between the definitions of health/illness and interactional mechanisms should be a focus of efforts to improve health.

2 Symptoms of health and illness in system theory are seen as part of the interactional processes, not as a characteristic of the individual appearing exclusively in his body and inner world. The notion of the 'identified patient' (see below; Minuchin 1973) was an early expression of this idea.

In Section 3 we will discuss the important contribution which system theory has made to health sciences by developing the notion of the identified patient. Subsequently, we will indicate how system theory can contribute in different ways to the improvement of human relationships (Section 4). Finally, these considerations will lead us to argue in favour of a culture which will be more concerned with the way in which we deal with each other (Section 5).

2 SOCIAL RELATIONS AND THE SYSTEMS APPROACH

The General System Theory (GST) has been in development from the middle of the twentieth century. The GST is a branch of science which transcends other disciplines and in which attention is given to the changeability of reality, its coherence and mutual interference (Burghgraeve 1986). GST has been applied to many fields. In the non-biological field, its contributions concerning psychiatry, psychotherapy and human relationships in general are particularly relevant to the health sciences. A lot of research in this area is still exploratory and abstract. There is little comparative or empirical research using strict methodological rules (validity and reliability safeguards, use of tested scales, control groups, etc.). Nevertheless, the lack of empirical material should not prevent us from exploring the potential of GST.

Since its origin the GST has been subjected to two evolutional shifts which produced what we will call the first- and second-order cybernetics, as they are usually called in the literature:

1 *The first order.* A system is mainly seen as a pattern of information and communication (the Palo Alto group, Bateson 1972; Watzlawick *et al.* 1967; Haley 1976; Minuchin 1973; Satir 1972, etc.).

2 *The second order.* The system concept becomes completely dynamic. Here a system is seen as a process without essence (chaos theory and self-reflective view: Tomm 1987a, 1987b, 1988; Prigogine and Stengers 1983; Maturana and Varela 1988; Keeney 1982; von Glasersfeld 1984; Anderson and Goolishian 1990; von Foerster 1981; Golann 1988).

The founders of GST (von Bertalanffy 1968; The Society for General Systems Research; Wiener 1948; Parsons 1937, etc.) held to a fairly closed and static notion of system. This was abandoned by the first and second order. The first order has produced important contributions to the health sciences. It has produced the notion of the 'identified' patient and has provided insight into the role which paradoxical communication can play in the development of psychological and/or psychosomatic disorders. The notion of circular causality used in the first order was based on the assumption that entities might influence each other mutually, but only to a certain extent: there was a hierarchy between systems. This was changed in the second order. Now the structure which takes shape is not enforced by a hierarchy, but develops by means of the combined action of the elements. Thus unpredictability becomes a central characteristic. The creation of order is not static, but is part of a continuous process of change. A second crucial assumption in the second order is the impossibility to observe and act neutrally: he who describes is being described (Varela 1989) and instructive interaction emanating from a one-sided influencing (of a therapist, teacher, etc.) is impossible (Maturana and Varela 1988). A third, new element is the introduction of networks, more specifically the entire social dimension (Hare-Mustin and Marecek 1990), instead of the family as the main meaningful unit.

The initial impetus to the formation of theories about human relationships derived from the system paradigm was provided by Bateson and a number of collaborators at the Palo Alto Institute in California between 1965 and 1970. They described so-called 'pragmatic aspects of human communication' (see Watzlawick *et al.* 1967). Using these grammatical rules as premises (and not as 'facts') generates a framework by which communication can be more easily understood. This approach was later adopted by many authors and elaborated on with additional concepts from the first and second order (Madanes 1981; Selvini-Palazzoli 1980; Andolfi 1979; Hoffman 1981; Dell 1982). We will now describe this approach systematically on the basis of three basic premises: the difference between the inside and the outside, the stratification of communication, and patterns of relations. All three phenomena occur simultaneously and are continuously interrelated with each other. For descriptive purposes we will deal with them separately.

2.1 The inside and the outside

One of the best known propositions of the general system theory is: 'one cannot not influence people'. The GST assumes that only the outside can be observed: the body and behaviour. Intentions, thoughts and feelings belong to the inside: invisible to other people. The outside does not contain information in itself. Yet it always influences the observer, who forms *impressions*, which do not always correspond to the inside, to the intentions, thoughts and emotions the other person actually has. *How people deal with*

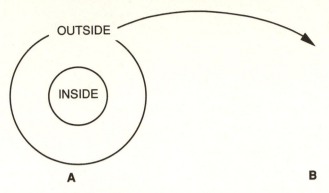

Figure 6.1 The influence of one's behaviour on another

each other, depends partly on the effects of the outside (Steens 1989). For example:

1 'I've really made a huge effort (my inside) but to the other person it looks as if I've been taking advantage (my outside, as observed by others).'

 'I thought the other person wanted to insult me (his outside as observed by me), but now it turns out he wanted to advise me (his inside).'

2 Patient comes to a GP with physical complaints. Judging from his knowledge and experience, the latter considers (inside of GP) them to be psychosomatic. The GP does two things: he talks to the patient about his way of living and refers him to an internist (outside of GP) in order to reassure the patient (inside). The patient's interpretation, based on the GP's outside, might be: 'He doesn't want to tell me how bad my condition is, so he refers me to a specialist.'

2.1.1 I do not influence

Since outsides always are perceptible, people always influence. Behaviour, according to the GST, always has a meaning (see Figure 6.1). For example:

1 He had decided to let her talk ('I do not influence her'). He does not realize how he is frowning, scratching his head, leaning backwards and forwards, etc. He does not realize either how she, willy-nilly, is select-ing and interpreting this information and how thus her behaviour towards him is being affected: 'He's having a bad day / he doesn't agree with what I'm saying, so I'll stop / I'll ask for his opinion.' Thinking that he presents no behaviour whatsoever and hence is of no influence, he might completely attribute her behaviour to herself : 'Now I'm

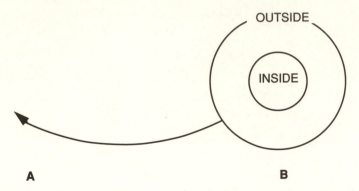

Figure 6.2 The influence of another's behaviour on self

giving her room and she still hardly says anything!'; 'She constantly needs my opinion'; 'Why is she being so offhand, I didn't do anything.'
2 'Shall we go to the movies?' Thinking his suggestion is free of any obligation, he might fail to realize that his large build, his heavy voice and his fixed stare, were turning this request into an obligation. Eventually, the irritation shown on the partner's face will have another meaning to him.

To be aware of one's own influence is extremely difficult. 'I have no influence, it has nothing to do with me, I'm only undergoing this experience, I'm only observing', are utterances or considerations which can arise when one is confronted with other people's behaviour. One is more aware of what one feels inside, than of what one does on the outside.

2.1.2 I am getting no influence

Since outsides are always perceptible, one is always getting influenced. To relate to one person, GST states, means being influenced by the latter's behaviour (see Figure 6.2). For example, the man in the former example may not realize he is of any influence, but he may not realize either that he is being influenced. Eventually, he will attribute his dissatisfaction to his personal, independent conclusions about her. What she does to him (unwillingly and unknowingly) that makes him think about her in the way he does, escapes him. He thinks that if she means anything to him, 'it's obviously because of what he made himself believe about her'. She herself has 'nothing to do with it'.

'I'm not being influenced, it's all my fault / I deserve all the credit, I'm on my own in this' are utterances or considerations which can arise from experiences, often negative ones.

It is a step further away from the awareness of influence. Only when one

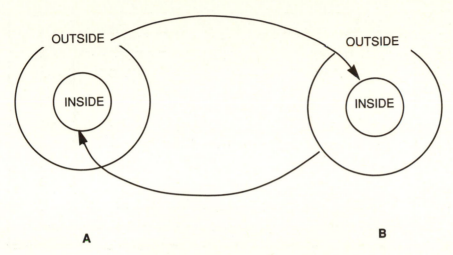

Figure 6.3 Simultaneous influence

constantly fails to see one's own influence, does the feeling of 'being of no importance' appear. One believes oneself to be out of the circuit of influence, 'a case apart', to be in a situation in which the environment plays no part (cf. illness).

2.1.3 'Who dunnit?'

Since outsides are perceptible, both partners in a relation are simultaneously influencing and being influenced. According to the GST, everyone is always both a transmitter and receiver of influence (see Figure 6.3).

Since influence occurs *simultaneously* on both sides, it is not possible to say 'who dunnit'. (Responsibility is a different issue: both for 'positive' and 'negative' behaviour, responsibility is attributed according to the ruling moral values.)

The problem is that *circularity* (the continuous, simultaneous, reciprocal interaction) cannot be represented simultaneously. Hence, in daily life, a *chronology* is usually practised:

- 'she smiled and he smiled back at her',
- 'he looked at her, she smiled and he smiled back',
- 'she turned round, he looked at her, she smiled, he smiled back at her',
- 'he passed, she turned round, he looked at her, she smiled, he smiled back at her', etc.

It is obvious that *the place* where one replaces circularity in favour of chronological representation is a *subjective* choice. A choice which is inevitably and immediately connected to a causal dimension:

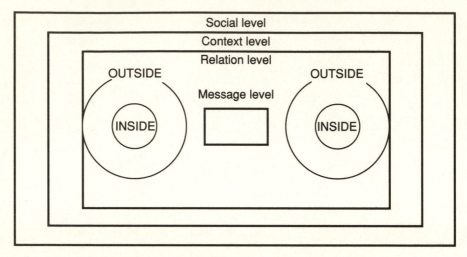

Figure 6.4 The stratification of communication

- 'while she turned round, he looked at her',
- 'she turned round while he was looking at her'.

People tend to ask 'who dunnit?' This reduces the interactional network to the contribution of only one person.

2.2 The stratification of communication

An important premise of the GST is the stratification of communication (Pearce and Crooner 1980). Communication takes place on at least four different levels:

(a) the message level;
(b) the relation level;
(c) the context level;
(d) the social level.

All of these four levels are of value at the *same* time, and are interrelated in a continuous, simultaneous and empirical way (see Figure 6.4).

2.2.1 *The message level*

The message level refers to 'what' is factually and verbally exchanged in information, goods, services, etc. The GST attaches a lot of importance to the differences of meaning and the misunderstandings which surround it. Amongst other things, the GST contributes a new perspective on the illusion of objective information.

People can *have the impression* that they understand each other, that they agree about an utterance, whereas they – invisibly to each other – give a (sometimes completely) different meaning to this utterance internally. For example:

- 'I won't be *long*.' 'We don't have *much* homework.' 'I'm *never* allowed to do anything.' 'I would like to see you more *often*'; 'If we want to keep seeing each other, we'll have to give each other *some freedom.*' What do 'long', 'much' and 'freedom' mean to the transmitter and receiver of the information?
- 'We'll never let each other down' (what does 'let each other down' mean?), 'We'll always give each other enough freedom.' (What does 'enough freedom' mean? what does it mean *now* and what will it mean *later*?)
- 'I'd really like to have a *nice* holiday'. What is a 'nice holiday': practising sports, rest, travel around, meet people, admire nature, acquire some culture, go as a couple, in a group with other couples, by car, by plane, via a travel agency, for a fortnight, three weeks, etc.?

2.2.2 The relation level

'What' is being exchanged in information (with all possible misunderstandings) is always linked with 'how'. 'How' implies intonation, volume, silences, eye contact, facial expressions, posture and all kinds of body language. According to the GST, 'how' indicates in which way one is supposed to understand the 'what' (e.g. the differences between 'you're mad' said with a smile, with a frown, and with wide-open eyes).

This non-verbal communication is called 'analogous' language by Watzlawick and others (1970). They emphasize its *multi-interpretability*. Thus, attention is drawn on this level too to the differences in meaning. Of course, a number of *shared* meanings exist, e.g. to burst out laughing is seldom seen as an expression of sorrow, fear, pain, etc. Sociologically it is particularly relevant to note that social hierarchies, or other differences between social groups and/or social categories, are often expressed via analogous language. Shifts in social position thus go together with shifts in analogous language, which can cause unexpected and unintended conflicts in the interaction. These shared meanings reinforce the illusion of mutual understanding of body language. An 'illusion' indeed, since according to GST analogous language is multi-interpretable and thus subject to misunderstandings; e.g. to burst out laughing could be: to have fun, to mock at, a hysterical attack, sympathy . . .

As mentioned earlier, the GST emphasizes how people influence each other continuously. They cannot stop relating to each other. Hence, in each kind of behaviour there is the potential to interpret this behaviour as a description merely about oneself (e.g. to scratch one's head: 'It itches') *and* simultaneously as a description of their relation ('I don't understand you') (Laing 1966). There are, of course, many other interpretations for both descriptions, e.g.:

- self-description: I'm having a rough time, I'm getting tired, I'm thinking.
- relational description: you're being difficult again, you're making me think (I hadn't thought of it this way), you're quite a character (one in a million).

But behaviour is present on both sides simultaneously: the behaviour of the partner in the relation also contains both a self description and a relational description. If the relation is to last, both partners will keep on checking whether they still mean anything to each other, whether they are still confirmed by the other partner in their descriptions. Even if they both agree to disagree, there is still a relation. According to this view, it is inherent to a relationship that people ask for confirmation of their descriptions. But the intention or the initiative to seek confirmation, determines the chronology of the experience.

For example: man and woman meet each other at the end of the day. Both would like to see their self-descriptions confirmed: 'I've worked hard today.' Even if they can both notice their self-description simultaneously, and want to acknowledge this simultaneously, they will have to make a big effort (work even harder) to make this happen *simultaneously*. Even if they fly into each other's arms simultaneously, they are still tied to a chronology of intentions: who starts walking first, who spreads his/her arms first, who says something about oneself first or asks something about the other?

Because of this phenomenon, people in a relation get caught up in a chronology of events very quickly: first he did . . . then I did . . . And because of this chronological representation, the occurrence of rejection and the loss of confirmation becomes more likely.

Let us analyse one situation at a given moment between A and B:

1   A: 'I've got a headache.'

Self-description: for example: I want to tell you I've got a headache, I'm tired, I've had a busy day, I'm getting old, I'm ill, . . .

Relational description: for example: you're boring, you haven't worked as hard, pamper me, leave me alone for a while, . . .

2   Without realizing it, B is selecting one of the two descriptions, e.g. the self-description with the interpretation of 'I'm tired'. At the same time he presents in his turn a behaviour which also contains a self- and a relational description. B says: 'Lie down for a while.'

Self-description: for example: I know what's good for you, I'm worried, I'll manage, . . .

Relational description: for example: you're tired, what do I care, you're always being difficult, . . .

3   Without realizing it, A is selecting one of the two descriptions of B. For example: A can select the description of B that links up with A's own description. A will then feel confirmed (A: 'I've got a headache' = 'I've had a busy day' – B: 'Lie down for a while' = 'I'm worried').

But A might also feel rejected by selecting another description (B: 'Lie down for a while' = 'you're always being difficult'). In that case A may very well reject this description as incorrect. This will cause A to show behaviour that B may see as a rejection of B's own description, which B, in turn, will reject as incorrect.

We recognize the principles of general system theory: there is always a behaviour, it always contains a self- and a relational description, it is inherent to a relation that people seek confirmation for their descriptions, the message of that description is always subject to a variety of interpretations, so the possibility that rejection becomes the main theme in the relation is very real. This usually happens unwillingly and unknowingly: everyone is preoccupied with seeking confirmation of their own descriptions (of oneself and the connection with the other) without realizing that the partner is doing exactly the same thing. Thus a spiral of rejection is created, by which people are less and less aware of their own influence and feel increasingly worse in their relationship. For example:

A: 'You're always criticizing me' (I want confirmation of my description, but I never get any).
B: 'You always have to be right' (I want confirmation of my description but you won't give it).

People are used to see each other as separate identities: whatever happens, happens either to one person or to another. When one is able to see identity as a continuous process of development which is inextricably linked to the process of human interactions, it becomes easier to realize what, among other things, occupies people: *to exist in the eyes of one another* (Mielants 1985).

2.2.3 The context level

'What' and 'how' get their meaning within a context (to say 'you're mad' with a smile: at the table or in bed, in the street or on the tram, alone or surrounded by friends . . .). The GST accentuates the difference in meanings which are given to contexts. Here also, there are a number of *shared* meanings, e.g. a disco is not seen as a place to trade (apart from those cases where drugs are dealt) or to spend the evening talking (unless one would want to use sign language to communicate). The underestimation of the importance of context leads to the illusion of equality, and thus unknown differences in meanings of contexts are subject to misunderstandings: a disco as a place where one has a good time, forms relations, relaxes, drives away loneliness, meets 'the beautiful people', etc. . . . A multitude of misunderstandings occur because one automatically assumes that the context in which one finds oneself is unambiguously (i.e. by everyone) *experienced in the same way*.

For example: a boy wants to hug his girlfriend on the stands of a football ground, but receives a rejection as a reaction. He might take this personally and not consider a different meaning of the context given by his girlfriend: 'People only hug each other on the pitch nowadays, not on the stands.'

2.2.4 The social level

The GST has been involved with the social aspect since its origin : 'Mind is social' (Bateson 1972). The previous examples show the importance of this fourth level: the interpersonal selection of socially formed meanings. This relates to the idea of the *shared self*: individual, but not private (Mattheeuws 1992). Thoughts and feelings are experienced individually but are generated during the social exchange of meaning. One is always male or female, parent or child, of a certain age, living in certain social conditions, consumer, pedestrian, etc. Through these I-pieces one belongs to groups: men / women, parents / children, the young / people in their forties, etc. There are social meanings attached to each group: a woman cannot / is allowed to / must . . . a good father cannot / is allowed to / must. These social meanings are continuously (re)formed in verbal and non-verbal dialogue between people. Misunderstandings arise when the expression (and simultaneously, creation) of social meanings is experienced as an inter- or intrapersonal message.

For example: a 27-year-old man can take the message 'Well? Are you still not married?' as a personal failure, rather than as an expression of a social meaning about marriage and the age at which this is done. 'I can't blame you, old boy, enjoy it while you can!' is an expression of another social meaning of the same subject which offers the possibility of a totally different personal interpretation.

People who share the same social meanings experience them as – what we might call – 'self-evidences'. They are not noticed, but they are passed on. Since not everyone shares the same dialogues, innumerable differences in self-evidences will exist about the same aspect of daily life. Because these things are experienced as self-evident, one is not likely to notice such differences. For example: 'love means involving each other in everything as much as possible' versus 'love means giving each other room'. Hence people can believe that they agree on the meaning of behaviour, and of the context in which this behaviour occurs, and of the groups to which one belongs, whereas in fact, this is not the case. This increases the likelihood of incorrect conclusions about oneself, one's partner and the relation.

For example: wife to husband: 'Please clear away those papers in the living room.' (A woman is responsible for the housekeeping. If we happen to have visitors, they'll look at me.) Husband may take this as a personal criticism: 'You're so fussy' (you don't care about those papers, their content is far more important than where they happen to be).

2.3 Relation patterns

The GST is associated with patterns: 'the pattern that connects' (Bateson 1972). From the very first moment of interaction, a *pattern* arises in the relationship: some forms of behaviour will be expected sooner and recognized easier than others. This gives the relationship a certain kind of *predictability*. Fluctuations can occur, of course, but these remain relatively predictable. A pattern also imposes a *'straitjacket'*. It is the counterpart of predictability. Partners in a relationship are forced by the pattern to repeat themselves.

There is nothing deterministic about a pattern: it can change at any moment. Still, relation patterns tend to persist. In order to explain the persistence of a pattern (a mental pattern, a social pattern), and of a pattern in relationships in particular, several influences are proposed:

- The pattern is generally beyond awareness: one acts accordingly without knowing.
- Even if – because of new information – one starts to recognize the pattern which influences the action, the old familiar behaviour is likely to be 'chosen' again. New behaviour has to be kept up for a long time before a new pattern establishes itself. People generally try out a new behaviour only a few times, become disappointed when the result fails to appear, and draw conclusions about themselves and/or the other person ('a difficult person', 'weak personality').
- Moreover, the partner in the relation will not be able to adapt immediately to the new jigsaw piece. Since partners *relate* to each other, he might – unconsciously – push his partner back in his old behaviour.
- Finally, and most essentially: the familiar way of dealing with each other has become a part of the relationship. A change in the pattern affects the way the partners see the relationship itself. New behaviour, however positive, is a threat to the familiarity of the relationship ('This is not us!').

3 THE CONTRIBUTION OF GST TO THE SOCIAL CONCEPT OF ILLNESS

In the 1940s, research into the aetiology of schizophrenia began in the United States. Partly founded on this research, the idea developed that serious pathology in an individual can be caused and sustained by what happens among the members of the system to which the individual belongs. Initially the attention was focused on the dyad mother–child (patient). At that time, the emphasis was on the pathogenic effects on the child (Bowen 1965). The term 'schizophrenogenous mother' dates back to that time. Further research, still mainly centred around schizophrenia, led to the observation that, within the context of a symbiotic mother–child relationship, fathers also played a more or less stereotypical role (Zuk and Rubinstein 1965). This role was hidden by the seemingly detached position the fathers had in these families. In short, the conviction grew that the cause of the illness was not to be found in the individual, or in the relation between said individual and a specific other member of the system, but in the whole system of interactions. Hence, not the individual, but the whole system was pathological.

This point of view has had far reaching consequences, both practically and theoretically (Rijnders 1988):

- Attention to the system as a whole.
- Attention to the system rules and their influence on behaviour (system rules are the unspoken and, to the members, unknown codes of behaviour which prevail in the system and which regulate the behaviour of the members of the system amongst themselves).
- A changing view on dichotomies such as illness–health, good–bad, etc.
- Emphasis on treating the system as a whole.

In 1973 Minuchin introduced the term 'identified patient'. He revealed on the one hand the relativity of the role of the patient, and on the other hand the environment (family, system) as a contributory factor. The concept of patient is thus given a new meaning, viz. that of a signal that indicates a dysfunctioning system.

In the field of psychosomatics, the GST has made important contributions (Minuchin *et al.* 1978; Haley 1959; Selvini-Palazzoli 1986). The GST makes a distinction between digital (verbal, literal) language and analogous (non-verbal) language. In order to function well, a well-functioning system of translation between both languages is necessary. Psychosomatic complaints, as well as other serious disorders like depression and anorexia nervosa, arise when experiences cannot be expressed digitally (literally and unambiguously), because of the rules of relationship. These are the unspoken regulations which indicate what is allowed and what is not. Experiences can then only be expressed analogously, via body language. This language is multi-interpretable by definition. On one hand this can lead to psychosomatic complaints; on the other hand, an excess of analogous language is very likely to be misunderstood or not understood, or to cause impotence or alienation. The latter can also increase illness. Estimates by both GPs and researchers about the percentage of psychosomatic complaints with which GPs deal on a daily basis, vary from 5 per cent and 30 per cent (Verhaak 1986). Estimates vary between 5 per cent and 90 per cent (Sanavro 1982) for the prevalence of psychosomatic complaints. This may suggest that numerous experiences are expressed in an exclusively analogous way; in other words, one cannot openly show what bothers one or what one feels. This deficiency leads to high medical consumption by patients who are obviously suffering, but who at the same time, as 'identified patients', show the dysfunctioning of the whole system.

4   SOCIAL RELATIONS AND HEALTH PROMOTION

One can propose a number of ways in which insight into human relationships can be improved in order to promote health. Educational institutions in particular, are of great importance for promoting and teaching this insight (Mielants 1991). Training is necessary to become accustomed to the complexity of interaction processes. We will use the same structure as in Section 2.

4.1 The use of feedback

By implementing the feedback principle to the understanding of inside/outside, the GST can make an important contribution to health promotion.

4.1.1 From intention to effect

International research on schizophrenic patients has shown how the expressed emotions of the members of the family significantly influence the outcome of the illness (Leff 1976; Leff and Vaughn 1981). Also, the influence of the verbal/non-verbal expressions of health-care workers on the pathological process has been recognized. Thus patients are influenced not by intentions but by actions. Health can be promoted by being aware of the effect we have on one another. Unfortunately one cannot, in advance and independently, predict which effect one's behaviour has on other people. Feedback can help: what do I get back from the other person which can show me the effect of my behaviour on this person?

4.1.2 From prejudices to checked impressions

One of the most important contributions of GST to health promotion is the notion of the preconceived frame of reference (Korzybski 1941).

Impressions of the other person are based upon subjective criteria concerning his or her behaviour, their appearance as a person and on the selectivity of observation. These prejudices lead to mistreatment unless they are recognized as impressions, and checked by practising feedback.

4.1.3 From chronology to circular causality

The more serious the disease, or the more extreme the behaviour, the more difficult it is to see the connection between behaviour or illness and spirals of influence. Nevertheless, these connections are always present, according to the general systems theory. The alcoholic, for example, is only part of the system (see Section 3) as is the suffering partner. Both have influence on each other. It is not easy to keep on realizing that individual experiences are only a fraction of the process of mutual influence. Feedback alone cannot undo labelling (e.g. towards illness) of the individual, but it can sometimes make it possible to uncover this mutual interconnection.

4.2 Coping with manipulation

However clear and exhaustive one believes oneself to be, it is always possible for information to be understood in a different way. Recursive questioning (i.e. using feedback; see Selvini–Palazzoli et al. 1980) is already being taught in health care, especially to GPs in training. In contrast, the holding back, twisting, obscuring or selecting of information persists because of the strength of habit. Authenticity and hence health can be promoted on both sides of the interaction by using recursive questions. The teaching of this method and the cultivation of the attitude, not only in health-care circles, but also in daily life, is an important aspect of health promotion.

4.3 Avoiding paradoxical communication and double binds

The stratification of communication (Bateson 1972; Cronen 1987), the differences between inside and outside, between the message and the relational level and between analogous and digital language, make it very likely that there will be uncertainty, confusion and misunderstandings in relationships. Sometimes an action or intention will thus have the opposite effect; the more one tries to achieve a goal such as rest, relaxation, appreciation, the more one might achieve the opposite. A contrary effect like that, is called paradoxical. Paradoxes and paradoxical communication have a negative effect on health (Evers and Rijnders 1982; Rijnders 1987). Paradoxes lead to confusion, turmoil, self-doubt and doubt of others, alienation and depersonalization. According to important systems theoreticians (Bateson 1972; Haley 1959; Selvini–Palazzoli and Viaro 1988), paradoxes and paradoxical communications are at the basis of so-called double binds. These are patterns of communication in which contradictory demands are made simultaneously. Best known is the so-called 'Be spontaneous' paradox. Systems in which double binds occur are frequently, according to these authors, a breeding ground for psychosomatic disorders and psychoses. Health is not helped by paradoxical communication and double binds. Health is helped by openness and clarity in communication. Both stratificated patterns of communication and the essential ambiguity of communication are inherent to life. Therefore, harmful developments can only be avoided when people have relational skills, like the giving of and asking for feedback, and when the systems in which they function 'allow' sufficient relativity.

4.4 Avoiding mutual rejection

Mutual rejection leads to negative effects in the experience of oneself or/and the other. Health can be promoted by interpreting rejection not as an individual, but as an interactional phenomenon: the simultaneous reciprocal demand for recognition. From this point of view a reproach on either side can be identified as a rejected (not noticed) call for recognition, and inexplicable behaviour as a feedback-sign of the same thing.

4.5 Breaking the illusion of equality

An unnoticed difference in the meaning of context automatically leads to a different interpretation of behaviour. This can cause conflict, confusion or labelling. People judge themselves and others without knowledge of the difference in meanings of context. Health can thus be promoted by using the experience of failure to understand or to be understood, as a warning signal: a warning of the possible lack on one or both sides of tuning in sufficiently to the difference in meanings of context.

4.6 Recognizing self-evidences

Health promotion means acquiring insight and knowledge about self-evidences in society within the different groups people can belong to. One is not inclined to invest in another person if one is experiencing negative behaviour as arising from the person. But if one can see people belonging to different groups behaving differently because of different self-evidences another (positive) interpretation of the behaviour might arise. From that moment on, one is in a realm of simply different points of view (self-evidences), rather than different labels (stupid? weird? sick?) and this improves health.

4.7 Avoiding labelling

People tend to attribute recurrent behaviour to fixed elements in a person showing this behaviour. As was stated in Section 4.6, one is not inclined to invest in this person whenever the recurrent behaviour is experienced as negative. Thus health can be promoted by getting hold of the idea of patterns of relationships. Consequently, there will be less of a tendency to link recurrent behaviour to individual virtues/vices. At the same time relational experiences may be seen as the consequence of the patterns typical of all relationships, rather than of a specific personal relationship.

5 CULTURE AND SOCIAL RELATIONSHIPS

People not only relate to one another. According to GST, they influence and are being influenced by larger entities, such as the society they are continuously recreating. As a consequence, the impact of relational knowledge, insight and skills, as described above, depends on the meaning it is given within society. On the other hand, the creation of interactional realities in dialogues between people, influence the societal culture. We would like to argue in favour of a cultural evolution in which people will be brought up to achieve more competence in understanding interactional processes. The insights developed by GST can contribute to this.

Insight into the circular and process-like nature of interactions can develop awareness of the interconnectedness of people, in contrast to the lack of solidarity felt nowadays. 'It has nothing to do with me' can become 'What I do or don't do cannot be separate from others.'

The use of feedback can stimulate people to let go their prejudices. A good example of such a closed vision is the classification systems in health care where the inside of people is being judged by standardized interpretations of their outside.

Insight into the phenomenon of mutual rejection can on one hand promote the managing of conflicts and human tolerance, and on the other hand help to recognize and stand up to unwanted psychological pressure and abuse of power. Insight into social relations can increase people's freedom of choice to decide how they want to live their lives.

In society today we can see a declining readiness to give thought to other people, and prejudices may be on the increase (e.g. the rise of racism and neo-fascism in Europe). Human values are no longer sufficiently visible. Developing knowledge concerning the existence and variety of 'self-evidences' can help to change this. Understanding differences in self-evidences makes it possible to put absolute attitudes about one's fellow man into perspective and to use negative impressions more as a sign: a sign that one does not quite realize which thoughts, feelings, intentions and views this fellow human being possesses. It also means paying attention to what seems self-evident: this encourages asking oneself what one does and why one does it. This may help to put one's own values into perspective. Knowledge and awareness of the existence of self-evidences will bring a heightening of mutual understanding and a reduction in discrimination and racism, and on the other hand, a renewed awareness of values in the society of tomorrow. All of which can contribute to the maintaining of a healthy and democratic society.

6   CONCLUSION

Continued research into the GST approach to health problems could be rewarding. A number of theoretical problems still require a solution. In particular, the theoretical insights which were developed in the second order need to be applied to the empirical domain of human relations. This, however, should not stop the world of research from making attempts in this direction. We can expect results from this type of research which may be of great use in prevention at the psychological and psychosomatic level. It is also essential to create a better empirical foundation, after the pioneering phase of the trend-setters. It can bring about theoretical refinement, and a selection between meaningful and over-speculative insights.

System theory proposes a dynamic, interactional vision of health and health promotion. Health promotion can be managed in at least four ways:

1   Gaining insight in the circularity of human interactions.
2   Realizing that interpretation of behaviour (both of one's own behaviour by another person and vice versa) depends on so many variables, that misunderstandings are daily occurrences.
3   Realizing that only feedback offers a solution to this: intentions are invisible, one cannot look into other people's insides. Feedback means: checking the effect on both sides, rather than living with presumptions.
4   Realizing that people always take part in a dialogue of social meanings. And practising to recognize these meanings in order to avoid inter- and intrapersonal misunderstandings as much as possible.

Apart from health-promoting activities which are aimed at the individual (medicine, counselling, psychotherapy), we argue in favour of prevention. To us, this means creating possibilities which effortlessly present and teach the aforementioned health-promoting activities, which are based on GST insights. Relational, health-promoting skills, should receive a prominent place in education as a whole.

REFERENCES

Anderson, H. and Goolishian, H. A. (1990) 'Beyond cybernetics', *Family Process* 29: 157–63.

Andolfi, M. (1979) *Family Therapy: An International Approach*, New York: Plenum Press.

Bateson, G. (1972) *Steps to an Ecology of Mind*, New York: Ballantine Books.

Bertalanffy, L. von (1968) *General Systems Theory. Foundation, Development, Application*, New York: Braziller.

Boszormenyi-Nagy, I. and Spark, G. (1973) *Invisible Loyalties: Reciprocity in Intergenerational Family Therapy*, Hagerstown: Harper & Row.

Bowen, M. (1965) 'Family psychotherapy with schizophrenia in the hospital and in private practice', in I. Boszormenyi-Nagy and J. L. Framo (eds), *Intensive Family Therapy*, New York: Harper & Row.

Burghgraeve, P. (1986) 'Epistemologie en systeemtheorie' (Epistemology and system theory), in D. Baert and A. Mattheeuws (eds), 'Nieuwe accenten in de systeem-theorie' (New accents in system theory), Antwerp: VStP, pp. 38–54.

Capra, F. (1986) 'Wholeness and Health', in *Holistic Medicine*, vol. 1, pp. 145–59.

Cronen, V. (1987) 'Het individu vanuit een systeem-theoretisch perspektief' (The individual from a system-theoretical point of view), *Systeemtheoretisch Bulletin*, 5(3): 167.

Dell, P. F. (1982) 'Beyond homeostasis: toward a concept of coherence', *Family Process* 21: 21–41.

Evers, R. and Rijnders, P. B. N. (1982) *Paradoxale benadering* (Paradoxical approach), Alphen a.d. Rijn: Samson.

Fauconnier, G. (1981) *Algemene Communicatietheorie* (General Communication Theory), Utrecht/Antwerpen: Het Spectrum.

Foerster, H. von (1981) *Observing Systems*, Seaside, CA: Intersystems Publications.

Glasersfeld, E. von (1984) 'An introduction to radical constructivism', in P. Watzlawick (ed.), *The Invented Reality*, New York: Norton.

Golann, S. (1988) 'On second order family therapy', *Family Process* 27: 51–65.

Haley, J. (1959) 'The family of the schizofrenic: a model system', *Journal Nerv. Ment. Dis.* 129(4): 357–74.

—— (1976) 'A history of a research project', in C. Sluzki and D. Ransome (eds), *The Double Bind*, New York: Greene & Stratton.

—— (1981) 'An interactional explanation of hypnosis', in D. D. Jackson (ed.), *Therapy, Communication and Change: Human Communication*, Palo Alto: Science and Behavior Books, Inc.

Hare-Mustin, R. T. and Marecek, J. (1990) *Making a Difference: Psychology and the Construction of Gender*, New Haven: Yale University Press.

Hoffman, L. (1981) *Foundations of Family Therapy*, New York: Basic Books.

Jackson, D. D. (ed.) (1968) *Therapy, Communication and Change: Human Communication*, Palo Alto: Science and Behavior Books, Inc.

Keeney, B. (1982) 'What is an Epistemology of Family Therapy?', *Family Process* 21(2): 153–68.

Koestler, A. (1979) *Janus – A Summing Up*, New York: Vintage Books, New York.

Korzybski, A. (1941) *Science and Sanity*, New York: Science Press.

Laing, R. D., Phillipson, H. and Lee, A. R. (1966) *Interpersonal Perception: A Theory and a Method for Research*, London: Tavistock Publ.

Leff, J. P. (1976) 'Schizophrenia and sensitivity to the family environment', *Schizophrenia Bulletin* 2: 566–74.

Leff, J. P. and Vaughn, C. (1981) 'The role of maintenance therapy and relatives' expressed emotion in relapse of schizophrenia', *British Journal of Psychiatry* 139: 102–4.

Luhmann, N. (1985) *Soziale Systemen – Grundriss einer algemeinen theorie*, Frankfurt am Main: Suhrkamp.

Madanes, C. (1981) *Strategic Family Therapy*, San Francisco: Jossey Bass.

Mattheeuws, A. (1977) Systeembenadering en Kommunikatietheorieëm (System approach and communication theories), *Leren en leven met groepen*, looseleaf edition, Samsom, p. 4410, Brussels: Alphen & Rijn.

—— (1992) Personal communication.

Mattheeuws, A. and Steens, R. (1977) 'Weerbaarheid en wapens: een weerbaarheid-straining' (Assertiveness and weapons: a training in assertiveness), *Leren en leven met groepen*, Brussels: Alphen A. D. Rÿn: Samson: 4410.

Maturana, H. R. and Varela, F. J. (1988) *The Tree of Knowledge*, Boston & London: Shambhala.

Mielants, P. (1985) 'Door de ogen van de ander: een systeemtheoretische benadering van zelfdoding' (Through the eyes of the other: a system theoretical approach), *Ethische aanspreekbaarheid* VVI Congres 1985, vol. 1: 124–31.

—— (1991) *Samen-zijn* (To be Together, a Textbook of Human Relationships), Brussel: ARGO.

Minuchin, S. (1973) *Families and Family Therapy: A structural approach*, Cambridge: Harvard University Press.

Minuchin, S., Rosmann, B. L. and Baker, L. (1978) *Psychosomatic Families: Anorexia Nervosa in Context*, Cambridge: Harvard University Press.

Oomkes, F. R. (1986) *Communicatieleer. Een inleiding* (Communication Theory. An Introduction), Amsterdam: Boom Meppel.

Oost, P. van (1977) 'Assertiviteitstraining: evolutie en relationele implicaties' (Training in assertiveness: evolution and relational implications), *Leren en leven met groepen*, losbladige uitgave, Samsom.

Parsons, T. (1937) *The Structure of Social Action*, New York: McGraw Hill.

Pearce, W. B. and Cronen, V. E. (1980) *Communication, Action and Meaning*, New York: Praeger Press.

Prigogine, I. and Stengers, I. (1983) *Order out of Chaos*, London: Wiley.

Rijnders, P. B. N. (1987) 'Een systeemtheoretische visie op psychologische theorieën' (A systemtheoretical vision upon psychological theories), in A. Mattheeuws and D. Baert (eds), *Psychotherapie, Psychologische theorieën, Systeemtheorie*, Antwerp: VStP.

—— (1988) 'Psychotherapie als raakpunt tussen persoonlijke en sociale waarheden' (Psychotherapy as a point of contact between personal and social truths), *Bulletin* 6: 218–33.

Sanavro, F. L. (1982) *Probleempatiënten in de huisartsenpraktijk* (Problem Patients in Family Practice), diss., Utrecht: NHI (Dutch Institute for Family Medicine).

Satir, V. (1972) *Peoplemaking*, Palo Alto: Science and Behavior Books, Inc.

Selvini-Palazzoli, M. (1986) 'Towards a general model of psychotic family games', *Journal of Marital and Family Therapy*, 12: 339–49.

Selvini-Palazzoli, M. and Viaro, M. (1988) 'The anorectic process in the family', *Family Process* 27: 129–48.

Selvini-Palazzoli, M., Boscolo, L., Cecchin, G. and Prata, G. (1980) 'Hypothesizing – circularity – neutrality', *Family Process* 19: 3–12.

Smith, M. (1975) *If I Say No, I Feel Guilty*, Baarn, Netherlands: Ambo.

Steens, R. (1989) 'Beïnvloeding' (Influence), *Samenspraak 72*, Hove: Inter-aktieakademie (Report of a conference).

Tomm, K. (1987a) 'Interventive interviewing: Part I. Strategizing as a fourth guide-line for the therapist', *Family Process* 26: 3–13.

Tomm, K. (1987b) 'Interventive interviewing: Part II. Reflexive questioning as a means to enable self-healing', *Family Process* 26: 167–83.

—— (1988) 'Interventive interviewing: Part III. Intending to ask circular, strategic or reflexive questions?', *Family Process* 27: 1–15.

Varela, F. J. (1989) 'Reflections on the circulation of the concepts between a biology of cognition and system family therapy', *Family Process* 28: 15–24.

Verhaak, P. F. M. (1986) *Interpretatie en behandeling van psychosociale klachten in de huisartsenprakijk* (Interpretation and Treatment of Psychosocial Complaints in Family Practice), diss., Rotterdam: Erasmus University.

Watzlawick, P. (1976) *How Real is 'Real'?* New York: Random House.

—— (1978) *The Language of Change*, New York: Basic Books.

Watzlawick, P., Helmick Beavin, J. and Jackson, D. D. (1967) *Pragmatics of Human Communications*, New York: Norton.

Wiener, N. (1948) *Cybernetics or Control and Communication in the Animal and the Machine*, New York: Wiley.

Wynne, L., McDaniel, S. H. and Weber, T. T. (1987) 'Professional politics and the concepts of family therapy, family consultation and system consultation', *Family Process* 26: 153–66.

Zuk, G. H. and Rubenstein, D. (1965) 'A review of concepts in the study and treatment of families of schizophrenics', in I. Boszormenyi-Nagi and J. L. Framo (eds), *Intensive Family Therapy. Theoretical and Practical Aspects*, New York: Harper & Row.

7

THE IMPACT OF NON-ORTHODOX MEDICINE ON OUR CONCEPTS OF HEALTH

Stephen Fulder

1 INTRODUCTION

Conventional medicine can be said to have been the dominant medical system in the post-industrial world for not much more than 150 years. Prior to that, although differential diagnosis was the main diagnostic method used, physicians had to compete with traditional practitioners and bonesetters, and apothecaries with Galenicists. Traditional and ancient medical concepts such as the four humours, the elements, the *vis medicatrix naturae*, and crasis/dyscrasis (i.e. that health is based on inner and outer balance) went out of fashion only during the early part of the last century (Rosenberg 1977).

In its short history, modern medicine has proved to be so apparently effective, and so well adapted to the industrial worldview that the impression arose that indigenous, ancient or traditional medicine had no validity and was virtually extinct. In fact, this was not so. It clearly existed in the East and the Third World, and in hiding in Western culture, where it took a defensive cultic posture in the face of modern medicine's self-confidence.

However, in the last 15 years there has been a radical renewal of interest in, and use of, traditional or alternative medicine. So much so, that conventional medicine sees itself actually threatened by an unscientific rival. Many are now visualizing a future for Western medicine in which it shares and competes for customers with alternative medicine, within a pluralistic medical system (Pietroni 1990).

The scale of the current use of alternative medicine is not always appreciated. Surveys have shown that in Europe roughly 1 out of 6 of the population have used alternative medicine (Oojendijk *et al.* 1980). In the UK approximately 13 per cent of the population seek alternative treatments (Taylor Nelson Ltd 1985) and this varies with presenting condition. Of British subjects with irritable bowel disease, 16 per cent seek alternative medical help (Smart *et al.* 1986). There are at least 15 million alternative medical consultations per year in the UK (Fulder and Monro 1985). Around

$2 billion per year is spent on alternative medicine in Europe (Frost and Sullivan Ltd 1987).

On a local level this can have a profound effect on the social attitude to medicine, and on the attitudes of physicians. For example, in the city of Oxford, England, there are as many acupuncturists as general practitioners. The former see the world of medicine in terms of Oriental universal energetic principles such as the '5 elements' and 'yin and yang', a view diametrically opposed to that of scientific medicine. This inevitably rubs off on the general practitioners, who not only read about these heterodox philosophies but are showing an increasing interest in learning the practices. Recent surveys have shown that three-quarters of British general practitioner trainees (Reilly 1983) and nearly half of those in practice (Anderson and Anderson 1987) wanted to learn one or more alternative medical techniques.

These developments may have profound implications on the study of health, which will be examined below. The implications arise from a growing familiarity within society of alternative explanatory systems concerning health and disease.

2 COMMON FEATURES OF NON-CONVENTIONAL MEDICINE

'Non-conventional medicine' is an aggregate term for a variety of ancient or traditional medical systems in their modern forms. They include the specialities in Figure 7.1, which are grouped according to broad similarity.

Some of the modalities in Figure 7.1 are complete medical systems with their own diagnostic and therapeutic methods based on a unique, global and self-consistent theory of health and disease (e.g. acupuncture, herbalism, homoeopathy and naturopathy). Others (such as the physical and manual therapies) are subsidiary techniques. Practitioners of these subsidiary methods do not consider themselves to be first-call primary-care practitioners.

Compatibility with conventional medical theory also varies. The physical therapies such as chiropractic, medical herbalism and to some extent naturopathy utilize essentially conventional diagnosis together with concepts of disease which are different from, but understandable by, conventional science. For example chiropractic is based on conventional anatomy and physiology, but extends knowledge of the pathogenesis and treatment of musculoskeletal problems (such as 'adhesions' and 'subluxations') into subtle areas that are regarded as invisible and unproven by conventional medicine. Herbal medicine recognizes and uses conventional descriptions of disease such as eczema. However, it chooses medicines that affect the supposed deeper imbalances (e.g. allergies originating from inadequacies in the liver) as well as treatments that attempt to restore proper local tissue function. On the other hand homoeopathy, naturopathy and acupuncture

Ethnic medical systems	Acupuncture and Chinese medicine Ayurveda and Unani medicine
Manual Therapies	Reflexology Chiropractic Osteopathy Alexander technique Massage therapy
Therapies for 'mind–body'	Hypnotherapy Psychic healing Radionics Creative therapies Anthroposophical medicine
Nature-cure therapies	Naturopathy Hygienic methods
Non-allopathic medicinal systems	Homoeopathy Herbalism

Figure 7.1 Therapeutic modalities of complementary medicine

involve radically different concepts of disease which are not easily translatable to or compatible with scientific medicine. There have not been any successful attempts to map these systems on to conventional medicine or vice versa. Even in China, where strenuous efforts have been made, there is still no agreement on whether acupuncture meridians or Oriental viscera such as 'kidney' correspond to any anatomical structures that we know (Rosenthal 1981).

The alternative therapies are based on ancient discoveries beginning with shamanistic and folk medicine, which gradually evolved into complete systems with unique theory and practice. Acupuncture can be traced back, via bone needles, many thousands of years. Naturopathy and herbalism go back to Greek and Egyptian medicine. Their philosophical context was essentially Hippocratic and Galenical, from where they derived their concepts on health. Health is seen as a correct balance or tuning of body processes. The constitution of the individual is his essential background. Inputs of foods, drinks, remedies, experiences, weather, or accidental events have different results with different constitutions. Thus, for example, people with cold constitution need 'hot' exercise and 'hot' remedies and foods such as ginger so as to avoid cold sclerosing diseases. The Hippocratic approach stresses naturopathic methods to detoxify the body so as to allow natural disease resistance. Galen synthesized the Hippocratic approach with the findings of the empiricists who believed, much like modern medicine, in a remedy for each type of disease or maladjustment. He was essentially eclectic overlaying Hippocratic theory of balance, the Pythagorean concept of the elements Earth, Air, Fire and Water (which are actually metaphors for solidity or heaviness, lightness and movement, heat and energy, cohesiveness and moistness), with a multitude of specific remedies and well-tried procedures. There are many similarities in Greek and Oriental medicine, for example in methods of sensitive evaluation of the inner state of balance of the constitution – both Galen and oriental medicine used qualitative pulse measurements. Galen stated he could measure 100 different types of pulse quality in the body, each with its own message as to the state of the interior.

There have been considerable changes in all the alternative systems of medicine, with much theory and practice being forgotten, some rediscovered, and new ones added. Homoeopathy, for example, is a system that stimulates the restoration of health using doses of the medicinal agent which causes reactions similar to the disease symptoms. This principle was in fact, used on the island of Kos in the prime era of Greek healing. However the phenomenon of microdose, and succussion (shaking) of medicines, is new. Today, alternative medicine is enjoying a flourishing renaissance in which technology is enriching certain modalities, particularly acupuncture. In its new form alternative medicine is still young, and society itself is not yet adjusted to it. Patients are not sure which therapy to use for which health problem. There are confusions and gaps in the proper training of practitioners and little official supervision over standards. Standards themselves vary from excellent to very poor. In addition, rate of increase is faster than society's ability to absorb it. Estimates put the rate of increase of trained practitioners in the UK to be 11 per cent (Fulder and Monro 1985) and the rate of increase in consultations to be 15 per cent (Davies 1984). This is several times greater than that of general practitioners. Neither at governmental, supergovernmental (EEC/WHO), local (local health and hospital management committees), administrative levels or in health insurance, is there sufficient understanding of this historic change in public choice.

3 CHARACTERISTICS OF ALTERNATIVE MEDICINE RELEVANT TO ITS DEFINITION OF HEALTH

3.1 Self-healing

A fundamental position of alternative therapies is that they attempt to elicit the self-healing capacities of the individual, and co-operate with them in arriving at a complete cure. The individual is endowed with a basic vitality and adaptability. If this natural resistance is utilized it grows stronger and will serve to defend the individual. Symptoms are not identified with a state of ill-health, but are only a manifestation of it. If they are attacked and removed, as in modern medicine, it not only cheats the body of a useful learning experience, but also of the state of complete health. This may damage basic vitality by suppressing the disease, or leaving it to fester invisibly.

Symptoms may be treated in alternative medicine, especially if they become unbearable, but very often they are left to clear up by themselves as the body is helped to cure itself. The symptoms may actually be used as a guide or beacon to direct the therapy. For example the ebb and flow of the severity of the symptom eczema, is monitored so as to reach a complete inner cure of the sensitivity. It is not suppressed with anti-inflammatory remedies. If the symptoms of itching and inflammation are so widespread or severe that they are unbearable or dangerous, they will be ameliorated with, for example, local herbal salves.

3.2 Individuality

Each person is different and will react differently both to the state of disease and to the treatment. The treatment strategy is different for each person, depending on a very wide range of influences – constitution, life-style, biographical history, external influences and risks, psychological disposition, etc. In acupuncture, for example, two people with the same condition in conventional diagnosis (e.g. gastric ulcer) will have different oriental diagnoses (e.g. 'hot' or 'cold' type), and will have a completely different treatment at different points. Each person takes a different individual route to return to health, and the treatment responds to this. Note that inheritance is acknowledged as a set of genetic predispositions, a basic quality of the vitality, given by birth.

3.3 The many levels of a human being

Health and sickness work at many levels, and there is no barrier between them. Mind–Body–Spirit, Inner–Outer, Individual–Social–Cultural–Environmental are all included in the creation of health or its destruction. For example Oriental medicine sees anger as sometimes the result of over-active liver, or anxiety as damaging the blood (a fact we have only now come to realize in the west after multi-million dollar research programmes on behaviour and cardiovascular disease). On the positive side, Oriental medicine emphasizes that when the opposites are in balance the spirit, the 'shen', is given opportunity. It is this pre-Cartesian assumption which has given rise to the label 'holistic medicine', in which a patient's goals, personality, attitudes, relationships, family, environment, etc. are all included in the task of medicine. In alternative medicine there are no *a priori* divisions like that of a skin specialist or a psychiatrist. The divisions are in terms of preferred strategies of treatment rather than sections of the human organism and ecology.

3.4 The end point of treatment is defined contextually

Treatment continues until the patient is as well as he can be or wants to be. This state of health varies all the way from a critically ill patient who wishes to die with strength and dignity, to the well patient who seeks treatment to be super-well. There is no defined state of wellness where treatment must stop, as there is in modern medicine at the point where there are no more symptoms to attack. This is particularly the case with naturopathy and Oriental medicine, where much of the work is devoted to prophylaxis, that is, to increasing resistance to disease.

3.5 Alternative medicine is safe

The Hippocratic statement *primum non nocere*, first do no harm, has been compromised in modern medicine, and redescribed as 'risk versus ratio'. In alternative medicine the techniques are developed so as to be essentially in tune with the natural working of the human organism, and therefore are likely to be very safe. Though no manipulation of the human organism can be completely and universally harmless, research has demonstrated that unsuccessful patients of conventional medicine often feel harmed by it, while the failures of alternative medicine are at least not harmed (Boven *et al.* 1977).

3.6 Different areas of competence from modern medicine

Alternative medical practitioners treat the kinds of health problems that conventional practitioners find difficult to treat – including musculoskeletal problems, especially arthritis and back pain, headaches, allergies, non-specific symptoms, anxiety, addictions, insomnia, chronic infections, gastro-intestinal problems, and chronic degenerative diseases. In such cases the slower and deeper methods of alternative medicine which require more self-help and changes to the life-style may be more successful than the aggressive techniques of conventional medicine. On the other hand conventional techniques would be expected to be more successful in acute, tropical, epidemic and genetic diseases, and in injuries. There are signs of an early sorting-out of the areas of competence of conventional and non-conventional medicine by the exercise of the patient's choice as a consumer (Fulder 1989).

3.7 Patient involvement

The therapist requires more active involvement from the patient, who therefore feels he belongs to a partnership in which both parties have a job to do. He is not a passive subject to whom things are done by a detached professional. Instruction on life-style is a part of most alternative medical treatments.

3.8 Health assessment

Health is viewed, in the clinic and in the home, as a process without a beginning and end. In this alternative medicine is close to the position of folk medicine. It has a major role to play in the area between complete symptom-free well-being, and actual disease. This 'third state' or pre-disease state, is viewed as a lack of health needing proper attention. 'Why dig a well after you are thirsty?' states the classic *Yellow Emperors' Book of Internal Medicine*. In other words although acupuncture and other treatments would theoretically not be necessary for someone who is vital, resistant and in harmony with himself and his environment, they have much wider application than conventional medicine since they also deal with vulnerabilities. Thus much of the work of alternative medicine may be with essentially symptom-free people.

In conventional medicine there is no operational language to assess health itself, and various states of health are only describable in common language (e.g. 'well-adjusted', 'vital', 'energetic', 'glowing') whereas in alternative medicine the practitioner is able to diagnose different states of health by diagnostic signs. These show states of functioning of the main organ systems.

The state of health is described by the performance of the systems of energy. These systems are relative to the basic genetic energy. Thus in Chinese medicine health is expressed by the quality of the 'chi' and the organ systems. The energy should be balanced, spread in the right way around the body, full, resistant, long-lasting, etc. In other alternative systems this measurement is also made using different conceptual frameworks. Thus balance between elements or fluids is tested in herbalism and naturopathy. Health is a dynamic state where these qualities are well-adjusted, and heat, moisture, etc. are balanced and in their right proportion and location relative to the constitution of the individual. In addition there are measures of the quality of organs (skin elasticity, brightness of eyes, clarity of skin, warmth, firmness of body, lack of waterlogging, etc.) which are similar to folk descriptions but more systematic.

Although health is relativistic, it is known, in alternative medicine, to have some standards. Very often these standards are set by the therapist himself, who is required and assumed to be healthy. In Oriental medicine the speed of pulse and breathing used to be measured against that of the therapist himself. There was a general rule that health implied four pulse beats to every breath. But in addition further measures of pulse needed to be taken which were only comparable to the therapist's pulse in the first consultation.

In subsequent consultations a constitutional picture of the patient would be established which would then act as a reference point. For example a patient may naturally have a very deep and slow pulse. This is his state when healthy. The therapist will attempt to raise and lighten the pulse a little by means of acupuncture so as to adjust the patient towards even better health and reduce his natural vulnerability to cold diseases. In addition if the patient's pulse became too light, erratic or fast the therapist would know a disease was brewing and be able to take corrective action.

In addition alternative medicine has skills to measure health by means of voice, sexuality, movement, body rhythms, vitality, longevity, subjective feelings, mental state, spiritual state, etc. This again is based on individual constitution and acknowledges wide differences between individuals even though there may be some generally accepted trends.

4 A META-MODEL FOR COMPLEMENTARY MEDICINE

It is possible to summarize many of the features of complementary medicine in a descriptive, qualitative manner. The basic axioms or concepts have been given above. When put together a working model would look something like this.

Imagine that the human being moves through his personal time and space, which is mapped out by biological, psychological and biographical events. He is in constant dynamic relationship with his internal, familial, physical and social environment. This relationship is subject to continual change and distortion, and this requires constant adjustment. This is essentially a learning process for both body and mind. Some health-related behaviour is always required so as to harmonize this relationship.

If the disharmony increases beyond the capacity of this immunity, beyond the knowledge and ability of the body and mind to cope, it may be necessary to seek assistance from complementary (alternative) medicine. For example, if the seasons change and the person feels pain in the joints or the first signs of a viral fever. Alternative medicine functions:

(a) To monitor the extent of the disturbance or distortion in the psychobiological field.
(b) To understand the inner and outer constitutional picture of the person, and the special susceptibilities of the person to his environment (including the social environment).
(c) To evoke appropriate self-healing capacities.
(d) To provide remedies so as to restore health in an individualistic and holistic manner. These remedies are not limited to the body but may include mental and spiritual healing.
(e) To halt and reverse disease processes and repair damage.

If the distortion is too great, the damage too severe, the disease acute or highly infectious or life-threatening, the condition becomes more suitable for the stronger methods of conventional, allopathic medicine. In such cases the patient is usually referred by the complementary practitioner or self-referred to allopathic health-care systems.

5   IMPLICATIONS FOR A DESCRIPTION OF HEALTH

There is much that scientists and health professionals can learn from alternative medicine about defining and measuring health. Knowledge about health could be increased in the following ways:

1   *Depth*. Alternative medicine could provide much deeper diagnostic signs of the proper functioning of the human organism, for example the voice, spread of warmth, pulse, etc. could be read and utilized.
2   *Individuality*. The acceptance of variability and individuality would make health definitions more valid and realistic. The nature of individual variation can be learnt from alternative medicine.
3   *Constitution*. This essential basis of all medicine has been lost in conventional medicine and needs to be relearned.
4   *Process*. The change of health through time, in the absence of disease, is also a skill available in alternative-medicine.

6   IMPLICATIONS FOR RESEARCH ON HEALTH

Alternative medicine presents a considerable challenge to science. Is the state of health of an individual measurable scientifically? Are the outcomes of manipulations and techniques in alternative medicine measurable by conventional scientific methodology? In medical research the conventionally agreed methodology for assessing a changed clinical picture is the randomized double-blind controlled clinical trial (RCT). However, there are almost insuperable problems in applying this to alternative medicine.

To begin with there is no agreed diagnosis. A patient may arrive with a 'kidney yin' deficiency, be treated and discharged, without a conventional doctor being aware of any initial diagnosis or any change. Deeper questioning may have revealed signs relating to mood, sexuality, activity, sleep, etc., but still, on discharge the patient will seem medically the same. Furthermore, a defined Oriental diagnosis, such as the one above, will manifest differently in different people depending on their constitutional baseline. Without an agreed initial diagnosis, similarity between members of the trial group, and agreed final diagnosis an RCT is extremely difficult.

If traditional diagnosis is abandoned, and a group of, say, asthmatics were collected for a trial of alternative techniques the problems remain. The asthma may be a minor symptom in natural medicine, which will be treating something deeper, while it is the identifying symptom in conventional medicine. Further, each patient must be treated individually, so the treatment cannot be standardized. And life-style components are essential in the holistic treatment, so that it will often be hard to know if it is the treatment that is succeeding or something else. Added to which, is the fact that the interaction with the therapist is an essential part of treatment, which makes double-blind methodology a problem; and it is also hard to introduce the placebo, since the placebo effect is itself one of the tools of the therapist and therefore part of the treatment.

Some efforts have been made to test alternative medicine by controlled trials. Usually a defined conventional diagnostic group (e.g. hay fever sufferers) are treated by placebo or the relevant treatment (such as pollen) despite expecting varied responses for each person. Some positive results have emerged (Reilly et al. 1986) but much of the power of the study is lost for the reasons described above. For a review of current research on alternative medicine see Fulder (1989).

The situation has resulted in a stimulus to new research methodologies which are appropriate for the examination of natural medicine. Many of these methodologies may be of use in the scientific investigation of health and health practices. These methodologies tend to discard the either/or question of conventional methodology, which attempts to strip the investigated situation of its richness by limiting the parameters. Instead the new methodologies accept the complex situation and develop forms of holistic enquiry that are disciplined. Subjectivity and bias are acknowledged, included in the research situation, and explained as a picture of the reality. Questions concerning the feelings of a person, and often of the enquirer or researcher too, are not only legitimate but are an essential feature of this kind of investigation, without which it is hard to proceed. Some examples of new methods of research follow.

1 *Single case study designs*. A new effort has been made to revise single case studies (Macleod et al. 1986). The problems which contributed to the decline in this method are those of subjectivity, bias and lack of generality. These are acknowledged but efforts are made to bypass them. Thus a patient may be monitored over time using different treatments at different times including placebo (although this presents an ethical problem). The patient acts as his own control. The issue of subjectivity is partly overcome by having a review or second opinion by an independent researcher who has not seen the treatment regimen. A great deal of valuable micro-information can be obtained in this way about the outcome of specific treatments (Guyalt et al. 1986; Kazadin 1982).

2 *Co-operative inquiry*. Pioneered by Peter Reason, John Heron and Rowan, in the UK, this methodology formalizes the group process. The patient, the researcher, the statisticians, the therapists, are all part of a team working together to achieve defined research goals. Everyone in the team is both experimenter and subject. Criticism is encouraged as a balance to bias,

but subjectivity is as admissible as objectivity. This has led to the concepts of 'critical subjectivity' as a working method (Reason 1986). Knowledge is built of layers of experience, and conclusions are both qualitative and quantitative. This research methodology is very suitable for research in psychotherapy and in psychosomatic conditions, in effects of life-style or behaviour on treatment, etc. This methodology is described in Reason and Rowan (1981).

3 *Facet theory*. A new concept for measurement of complex interactions, especially designed for use in analysis of the patient/therapist interaction and its outcome, is facet theory. It is based on questionnaires each of which address one facet, or conceptual area. The facets are then plotted on a three-dimensional grid to examine the connectedness of various facets with each other and the significant relationships (Canter 1985). For example, in analysing patient satisfaction and attitude through interviews, the facets may be practitioner behaviour (questions, leading to explanations, leading to actions) on one axis, illness dimension (from physical to psychosocial) on another axis, causal attribution (cause and effect of illness) along a third axis, and degree of patient concern along a fourth axis. The questionnaire results are plotted in space to determine significant relationships. This method, developed at the University of Surrey, UK, has already provided new insights into the therapeutical relationship (see contribution by Canter and Nanke in this volume).

7 CONCLUSIONS

Alternative medicine is rapidly increasing in popularity. As its conceptual basis is prior to, and in many respects opposite to, conventional medicine, it provides a challenge which has not yet been accepted. Alternative medicine's concentration on healing the healthy as well as the sick, and its familiarity with the origins of disease on the ground of human life, give it special skills at understanding states of health. It sees health as a process which is intricately related to the way a constitution and personality develops through time within the matrix of all the influences and relationships that are experienced.

A science of health would benefit from including the dynamic and practical descriptions of health provided by alternative medicine in its brief. It will bring to it more meaningful ways of describing and measuring health, for example by including constitution and the quality of vitality. It will provide a disciplined background to holistic studies. It may also provide a rich source of ideas and techniques for health promotion, some of which (e.g. lower levels of fat in the diet, raised fibre, certain vitamins to lower cancer risk, the health risks of food additives) have already been acknowledged by conventional medicine.

Nevertheless different ways of assessing health interventions are urgently required. It is an intolerable situation that just one of naturopathy's self-care measures, that of lowering the intake of animal fat for cardiovascular health, should have to be tested using billions of dollars of research money and many years of investigation, yet the question is still not finally decided, so that official health advice on this question remains ambiguous. If more appropriate assessments of health measures were used it would be much easier to ascertain what is healthy and what is not, for each person. The assessment methods now being developed for alternative therapies may be able to assist in this goal.

REFERENCES

Aldridge, D. (1990) 'The delivery of health care alternatives', *J. Roy. Soc. Med.* 83: 179–82.

Anderson, E. and Anderson, P. (1987) 'General practitioners and alternative medicine', *J. Roy. Coll. Gen. Pract.* 37: 52–5.

Boven, R., Lupton, G., Najman, J., Payne, S., Sheehan, M. and Western, J. (1977) *Current Patients of Alternative Health Care – a Three City Study*. Appendix 8 of Report of the Committee of Inquiry into Chiropractic, Osteopathy, Homeopathy, and Naturopathy, Parliamentary Paper No. 102. Australia Government Publishers.

Canter, D. (1985) *Facet Theory: Approaches to Social Research*, Heidelberg: Springer-Verlag.

Davies, P. (1984) *Report: Trends in Complementary Medicine*, London: Institute of Complementary Medicine.

Frost and Sullivan Ltd (1987) *Alternative Medical Practices in Europe*, Report No. E874, Grosvenor Gardens, London: Sullivan House.

Fulder, S. (1989) *The Handbook of Complementary Medicine*, London: Coronet Books.

Fulder, S. and Monro, R. (1985) 'Complementary medicine in the United Kingdom: patients, practitioners and consultations', *Lancet* 2: 542–5.

Furnham, A. and Smith, C. (1988) 'Choosing alternative medicine needs a comparison of beliefs of patients visiting a GP and a homeopath', *Soc. Sci. Med.* 26: 685–9.

Guyalt, G., Sacket, D., Taylor, D. W., Chong, J., Roberts, R. and Pugley, S. (1986) 'Determining optimal therapy – randomized clinical trials in individual patients', *New Eng. J. Med.* 314: 887–902.

Kazadin A. E. (1982) *Single-Case Research Designs: Methods for Clinical and Applied Settings*, New York: Oxford University Press.

Macleod, R. S., Taylor, D. W., Cohen, Z. and Cullen, J. B. (1986) 'Single patient randomized clinical trial', *Lancet* 1: 726–8.

Oojendijk, W. T. M., Mackenbach, J. P. and Limberger, H. H. B. (1980) *What is Better?*, The Netherlands Institute of Preventive Medicine and the Technical Industrial Organization.

Pietroni, P. (1988) 'Alternative medicine', *R. Soc. Arts. J.* 136: 791–801.

—— (1990) *The Greening of Medicine*, London: Gollancz.

Reason, P. (1986) 'Innovative research techniques', *Complementary Medical Research* 1: 23–39.

Reason, P. and Rowan, J. (eds) (1981) *Human Inquiry: A Sourcebook of New Paradigm Research*, Chichester: John Wiley.

Reilly, D. (1983) 'Young doctors' views on alternative medicine', *Br. Med. J.* 287: 337–40.

Reilly, D. T., Taylor, M. A., McSharry, C., Aitchison, T. (1986) 'Is homeopathy a placebo response? Controlled trial of homeopathic potency with pollen in hay-fever as model', *Lancet* 2: 881–6.

Rosenberg, C. E. (1977) 'The therapeutic revolution: medicine, meaning and social change in the nineteenth century', *Perspectives in Biology and Medicine* 20: 485–506.

Rosenthal, M. M. (1981) 'Political process and the integration of traditional and western medicine in the People's Republic of China', *Social Science and Medicine* 15A: 599–613.

Smart, H. L., Mayberry, J. F. and Atkinson, M. (1986) 'Alternative medical consultations and remedies in patients with irritable bowel syndrome', *Gut* 27: 826–8.

Swayne, J. D. (1989) 'Survey of the use of homeopathic medicine in the UK health system', *J. Roy. Coll. Gen. Pract.* 39: 503–6.

Taylor Nelson Ltd (1985) *The Monitor Programme*, Surrey, UK: Epsom.

8

EPIDEMIOLOGY AND THE CRITICISM OF THE RISK-FACTOR APPROACH

Jens-Uwe Niehoff and Frank Schneider

1 INTRODUCTION

The central tenet of preventive medicine is regarded as the eradication, or at least avoidance, of *risk factors*. This requires the identification of those under risk for the purposes of monitoring, education or treatment. Therefore epidemiology is seen as the science of research on risk factors and the scientific basis for prevention.

However, this approach is subject to constant criticism, not so much on epidemiological or statistical grounds, but on theoretical and ethical grounds. For prevention aims to alter the way we live, and preventive strategies not only reflect knowledge about health and illness, but also social values, traditions, history, social and economic interests and beliefs. Critics of the risk-factor approach are therefore not against prevention. Rather they hold that preventive strategies can be much improved by a more appropriate model of the needs of the individual and society.

The following arguments summarize our critique of the risk-factor approach (Niehoff 1978).

1 Risk factors describe populations with different patterns of health problems, but do not explain the causes of the different patterns which are often connected to social inequalities.
2 Risk factors are frequently defined as an individual problem, which supports the tendency to blame the victim (Crawford 1977).
3 Risk factors tend to be seen as causes of disease, leading to the erroneous belief that avoidance of such causes will eliminate disease.
4 Risk factors support the medicalization of prevention rather than its socialization.
5 Risk factors tend to focus attention on the behaviour of individuals rather than the quality of the environment, living conditions and other problems of the society.
6 Risk factors support a paternalistic strategy of health education, and do

not encourage people to change unhealthy living conditions. This detracts from self-determination and individuality as social values.

This contribution will draw on selected arguments to show that the risk-factor approach narrows the concepts of prevention and epidemiology. It will not attempt ro review other theoretical and practical aspects of the risk-factor approach.

2 HEALTH AND DISEASE

Thinking about risk factors leads us inevitably to the question of our objective, namely health. But what is health? Apart from definitions of health, there is no satisfying explanation of what it is. The term 'health' is a theoretical construct, reflecting a biological, psychological and sociological context.

On the one hand health is seen as an ideal status to be followed by everyone. It is a quality of function and an ability to maintain homoeostasis in the midst of changes in environment and activities. This ability is the result of both inherited and self-care conditions. Thus health is seen as a social norm, and a question of life-style and self-responsibility. This view of health defines implicitly what health should be.

An alternative view sees health as varying in different individuals. In this view everyone has his or her own health, and specific health biography, which could include states of well-being, complaints, chronic illnesses, handicaps and disabilities. Health is seen as a process, part of the processes of growing up, of socialization and of ageing which are in constant inter-action with the physical and social environment.

Both these views influence methodological positions in epidemiology. The first compares reality with a vision or model of what a healthy human being should be like, which we should all follow. It reduces human beings to a prototype, and limits their variation. It leads to the measurement of health as an individual's absence of illness, or their characteristics, reactions, or ability to cope with work. Health is judged by demands, capacities, capabilities. It leads to the recommendations not to eat badly, not to smoke, not to drink, not to be overloaded by stress or work, not to drive too fast, to avoid environmental hazards, to reduce cholesterol and blood pressure, to exercise properly, and live in a nice house in the country. Health is incompatible with falling ill, being off work, and using too much medical services. It is also incompatible with being chronically ill, handicapped, in need of social assistance and old.

Health has, in this view, a moral dimension, illness being seen as a kind of punishment for not following a healthy way of life. Health is the natural state, synonymous with youth, power and beauty, and it is a duty to maintain it. This deeply influences the social security system, and indeed all the political positions and systems that deal with health.

It leads to an obvious problem: that if we follow that norm, almost everyone is excluded. Hardly anyone will be without risk factors that are the result of daily life, not to mention the 'risky' life of the handicapped, old or chronically ill. It is impossible to assess the number of the healthy and the sick within the population, for everyone has periods of health and illness in their lifetime. Further, the categories of health and sickness are themselves defined in practice not so much from theories of health as from calls on medical services. And medical services themselves depend on the general structure of services for the population. Thus even the assessment of health depends on social issues.

The second view of health is not determined by normative ideas about health but by the reality of people's life and their social needs. Health is seen as a process reflecting changes in culture, living conditions, etc. For that reason it is also a historical category, changing its nature along with social developments. It depends on factors like socio-economic conditions, education and profession, age and gender, working and environmental conditions, or the physical and mental state of the individual. In other words, everyone has *his/her own health*.

Research on health is therefore research on people's lives. The reality of health is the mirror of people's life and social values. Health cannot be defined, only a healthy person, who is, of course, almost impossible to define. As soon as we discuss health, we are forced to consider the differences and inequalities between people.

We can focus on three aspects of the problem.

(a) During an individual's life, deterministic models are less and less able to predict health. For example early death in a population equalizes health and medical needs. As life continues individual needs become more varied, their health state becomes more varied and they show more individuality. For this reason cohort type studies using deterministic models, are quite likely to fail. For the same reason deterministic models can often seriously mislead individuals in their life plans.

(b) All diseases are unequally distributed over the lifespan. The structure of health problems within the population is a function of the age structure. In particular, as we pass through life stages the stresses and types of complaints change. We cannot apply the same concepts of causality and prevention in childhood, as in adulthood and ageing. Change in the plans for health promotion becomes ever more important during the lifespan.

(c) The health biography of a cohort of people is extremely variable because of biological and social variation. The degree of social variation can change more easily than biological variation, and social inequalities can be overcome more easily than biological vulnerability. This suggests that we should differentiate inequalities in health from differences in health.

There are not only methodological and theoretical problems associated with the attempt to arrive at an operational definition of human health. There are also ethical problems. For example such a definition could be used to divide people into different groups according to their health qualities, to meet the needs of the labour market.

In summary, the term 'health' is a theoretical construct which is the aggregate of all states of health that exist. It should be considered in relation to a self-determined life, and personal values such as self-realization, contentment, and responsibility.

3   RISKS AND CAUSES

There has been a long and endless debate on the problem of causality, and it must be briefly addressed here as part of the understanding of risk factors. The usual view can be summarized by the axiom:

If X is identified as the cause of A then avoidance of X necessarily avoids A.

However this leads to the absurd conclusion that if we control all Xs we have no As: that is, total health is achieved by total control. It is forgotten that causality is not reality but an interpretation of it or a methodology. Since everything is connected to everything else, avoiding a cause, X, in practice, will produce a variety of unpredicted results. In other words, we have destroyed a holistic view of the world by looking for causality, and seek to rebuild it with the use of causality. In research this method will work, but only in so far as the cause and effect are very closely connected in time, and highly simplified models are used.

Causality is not defined by an agent, such as a bacterium, parasite or behaviour, but by the interaction between this agent and the object. Any change in the context of the cause can have unpredictable consequences on the object. Causes do not exist independent of results, and therefore X must be defined according to A. The problem is that we find it hard to define A. If A is a disease, we normally define it by grouping it into categories by similar characteristics, such as symptoms. But diseases exist in individuals, and to that extent, every disease is different. If the result is more or less undefinable, we have to conclude that the surrounding world of the individual is all potential 'cause' with unpredictable results. If we fail to appreciate this we can do real damage to health and the environment. For example we use antibiotics to kill bacteria that infect people. But this weakens the resistance of the person and increases the possiblities of bacterial resistance, which requires yet more antibiotics which are yet more toxic, and so on.

121

An alternative is a holistic view of the world. This is, in contrast to causality, not a method of observation but the results of observation. It takes into account the fact that everything is in relationship to everything else. The network of relationships is endless.

4   IF EVERYTHING IS A RISK, WHAT ACTION IS POSSIBLE?

This is a question that should be asked by professionals who design programmes to support health. It is certainly asked by individuals who often perceive that they have little chance of finding their way through the jungle of risk factors to meet the 'heroic' definition of a healthy life.

Everything is risky; a life of poverty, or of opulence; a boring, protected life; employment as well as unemployment; there is a risk of dying young, and of growing old; to live in the town or in the countryside; to live a natural life, as the developing countries, or in the centres of civilization; it is risky to be a woman and risky to be a man; risky to intervene in natural processes, and risky not to intervene.

Obviously the detection of risks is a sophisticated methodological problem, and the assessment of risks seems to be the main challenge in promoting health. Yet there are serious difficulties in the way of developing any criteria for assessing risks. The main one is that all diseases are unequally distributed within all populations. There are many reasons for this including:

(a) All health risks are themselves unequally distributed.
(b) The likelihood of people becoming exposed to sufficient risk to produce disease is also unequally distributed, and individualistic.
(c) The degree and kind of sensitivity of individuals to the same risk will also vary undefinably.

Thus in any population we find a certain pattern of diseases and risks at any one time. These patterns are the results of the structure of living conditions, of the state of the individual, and of the state of the society, including the distribution of labour, goods, education, housing, etc. In relation to socio-economics, there is a profound correlation between the social position and health, and health state is a sensitive indicator of socio-economic development.

Research on prevention should therefore focus on (a) the structure and dynamics of all the risks to health, and (b) the structure and dynamics of the socio-economic ability to cope with the risks. Since, as we have seen, everything is potentially a risk, the task is so difficult that researchers take the easiest way out: to examine only the risks that are in the control of individuals, and that individuals can alter. In this view health is only achievable by influencing the individual, and thus becomes a purely medical and educational matter. The conditions of life are left untouched.

We can conclude:

1   There are many ways to fight an illness. All such strategies need specific knowledge about causes and risks, but demand specific sociopolitical preconditions in order to succeed. At the same time these sociopolitical preconditions influence the theories of prevention and the corresponding research.
2   Knowledge of the causes of disease is not the only thing needed to prevent them. For life has endless risks. Indeed, risks can be regarded as essential in order to make life worth living, and are often identical with opportunities.
3   The kind of knowledge necessary to act on health depends on the social and political infrastructure of the society. The struggle for a more holistic view of causation usually leads to political conflict.
4   All attempts to properly deal with risks will create conflict with social interests. The subject of prevention is laden with values and interests which should be expressed by all concerned.

5   HEALTH AS A SOCIAL VALUE

It is often stated that health is the most important social value. It sounds a convincing ethical and moral position, and the programme to eradicate all the risks for an individual seems completely justified. Yet eradication of risks could turn out to be dangerous, and people might not always agree to it. For example the hungry would prefer food as a priority rather than lessening health risks. Consider:

- The same fact could be considered by some to be a risk, and others to be an opportunity, or both a risk and an opportunity.
- A risk could be dangerous not for the present but for a future generation.
- A risk might arise from historical or social developments.
- A risk could simply indicate a biological characteristic of an individual, such as fatness, or an appropriate cultural behaviour, such as going to a pub, or ethnic or religious tradition or values, such as circumcision, or social standards, working methods, etc.

Faced with the choices in life, it becomes rather hard to maintain the position that *health* is the most important value. Yet the risk-factor approach leads to a paternalistic health education which sets the criteria for true health and assumes that everyone must follow them. It may even set and control social norms, such as diet and the availability of certain foods, which force people to follow such recommendations.

Professionals set the criteria for assessing risks, and their assessment, as we have seen, is different from those of ordinary people. This is one area where

123

people may suffer from criteria which are set by science. There is a hidden social conflict here which is often resolved by: (1) following an authority (e.g. religion, ideology, or tradition); (2) following a scientific theory (e.g. about nature, health, growth); (3) allowing people to make their own decisions and determine their own criteria (e.g. by democratic social structures, economic independence, etc.). These approaches require some comment.

1 *The message of authority*. It may be 'modern' to blame authorities. But their role depends on historical circumstances and on the kinds of problems to be solved. Legislation has been extremely important in controlling large-scale health risks such as epidemics, by regulations in relation to hygiene, sanitation and vaccination, over the last 150 years. However since health is so interconnected with socio-economic development, it may be that the role of the state will increase rather than decrease in the future, and lead to over-regulation of society in the name of health.

2 *The message of science*. Science provides descriptions of reality, explanations, and then generalizations or theories. However, this process of collecting and ordering facts, for example about health, does not automatically qualify science to intervene in reality. Research on health risks is not so effectively tied to health as, for example, research on health promotion. In addition, prevention is about influencing the future. To influence the future one needs to predict the future, and in this respect science has attempted to become a contemporary Cassandra. There is not only a chance that, like Cassandra, those predicting a poor future (full of risk factors) will suffer, but also that society will be divided into a kind of scientific theocracy or dictatorship of those who predict the future and create programmes for the future, and those ruled by such theories. In such a social dictatorship, the media play the role of the guards, frightening people by worrisome health messages, and giving advice on how to behave.

3 *The message of self-determination*. Though this is a fashionable concept, it is unclear what it actually implies. In a democratic society it is obvious that self-determination is the way to determine strategies for prevention, and health agendas. People cannot always be subject to others' definition of 'the healthy way of life'. On the other hand, 'the people' means everybody in society, and there are interests expressed by different social groups. The slogan of self-determination may be a middle-class view, held by people who already have security. Perhaps the working class would be more concerned about working conditions and social insurance than about independence. It was the working class that demanded that the state control social and working conditions. However, with the collapse of Marxism, it seems that the future is not in the hands of the working class, and, in relation to prevention, it seems to be more and more in the hands of the middle classes. The current health movement is the result of the middle classes expressing social responsibility. In other words, self-determination is difficult as a

general strategy as it requires balance and compromise between different social interests.

In practice, holding the view that health risks must always be avoided, has opened up a market of health. Today, not only illness has its market, but health too. Although most people are healthy most of the time, if they see themselves subject to health risks they will buy drugs or goods which are advertised as helping to avoid such risks. This will have economic and regulatory consequences, and even the assessments of risks will be influenced by the interests of the health businesses. This is already happening: for example, the food industry uses the demon 'cholesterol' to sell its products and influence the type of foods on sale.

It is usual to assess a risk by measuring how well it predicts a disease. There are cases where a risk factor predicts an event very well, such as chronic alcoholism and liver pathology. However, in such cases the prediction usually works well with only a small proportion of those who have the disease, in this case liver problems. It is generally true that the better the risk as a predictor, the smaller fraction of the total incidence is predicted, and conversely, the lower the risk (and the poorer it is as predictor) the wider the prediction among the population. To make a real impact on disease with a large number of people, one usually deals with low risk levels, which requires years of mass education or intervention. A high risk level and good predictor in relation to heart disease would be high blood pressure, but this only affects a limited number of people. Diet is a risk factor with a lower prediction level, but may be applied to a broader part of the population; change to diet requires social action and control over a long period.

6 RISK FACTORS AND STRATEGIES IN PREVENTION, HEALTH PROMOTION AND EPIDEMIOLOGY

Prevention, health protection and health promotion are different concepts and involve different targets, systems of belief and responsibilities.

Prevention is the group of specific measures which prevent specific diseases, for example by vaccination or lowering blood pressure. This has certainly been successful in the past, but it is mostly targeted towards individuals without dealing with the basic causes of the health problem.

Health protection is a wider concept and focuses on broader goals such as creating healthy living, working and environmental conditions, providing social care for the elderly, the disabled and the chronically ill, and guaranteeing free medical services. Health protection is basically normative, affecting everyone in society. It is extremely important in insuring health – one only has to think of the results of the social regulation that drinking and sewage water must be kept separate.

Health promotion is a newer concept and it has come into operation in response to new social demands and conditions in post-industrial societies.

It is based on the understanding that health can only be considered in the context of the relationship between an individual and the social and natural environment. Health promotion seeks to improve health by focusing on living conditions within the framework of political and social relationships.

Health promotion as a social movement is more or less in the province of the intellectual middle class. In a sense the middle class has taken charge of social progress from the working class, using the opportunity to focus on their requirements for health. For example they are more concerned with preventing stress-related problems than the working class, and it shows how different agendas for health may be held by different social groups. Though health promotion may be strong in ideas, it may also be weak in power, since it is supported by this one sector.

The kinds of strategies envisaged for improvement of health reflect social developments. In the past health protection was completely identified with the struggle against inequality and poverty, and the power of the working class attempting to improve their conditions. It has indeed now been confirmed that one cannot improve health without developing society as a whole. However the actual social influences on health strategies have changed with time. Thus health promotion requires more responsibility from the individual, partly because in the modern world the individual has more power to influence and threaten life around him. One only has to consider the awesome responsibilities involved in the Chernobyl disaster.

It might be thought that responsibility for health is a question either of the individual or of the state, the two being essentially opposite. However, if man is a social animal, then the opposite of state responsibility may be social responsibility. The problem with social responsibility is that society is not homogeneous, and, as we have seen, health agendas are subjects of dispute between different social groups. The strategies of health promotion may be endangered by this kind of struggle.

The fact that health promotion is rooted in social life is pointed out by Salmon: 'The social context in which people live is also the context in which disease arises. The spread of diseases in specific groups can only be understood through precise and elaborate examination of their social organisation. . . . This requires accompanying qualitative methodologies such as the political economic approach.' (Salmon 1989: 48).

Although health promotion is a new social concept and challenge, the term arrived in the modern world before there was a proper understanding of the concept. Instead of developing new strategies and weaving them into the modern world and the interests and needs of its people, we get the impression that 'old' strategies of prevention and health protection received a new label. As Brown describes: 'Unfortunately, most health promotion programmes begin and end with such efforts targeted to individuals. Individuals may find it difficult to adopt healthful behaviours because their physical or social environments impose many obstacles to engaging in these actions. Indeed, rather than holding individuals reponsible for their life-style, we should recognise the important role that physical and social life play in shaping both personal behaviour and health status itself . . . it is important to keep in mind the limited power that individuals have over their lives' (Brown 1989).

Health promotion is directed at people in all their variety, and not at diseases or at abstract and ideal patterns of health. As new insights into the risks to health have emerged, and new technical opportunities arise for interfering in the health biography of people, questions concerning the ethics and wisdom of using this knowledge also arise. These questions need to be answered by the whole of society and not just its scientists.

With health as a matter of social policy, we can focus on four tasks or problems which are of major importance and concern in health promotion:

1 Extending the field of democracy so as to set strategies and tactics of health policy within society.
2 Increasing publicity about health in relation to working, living and environmental conditions.
3 Supporting the role of communal politics and groups in health promotion.
4 Reformulating health as a goal of social policy.

Reformulating strategies to improve health is not just a task for epidemi-ology and medicine, but for all the social sciences and the health sciences. To some extent epidemiology is a social science too, for it not only collects and describes risks. It reflects an aspect of social life, namely the cause of changes in the health status of a defined population. Differences in health within a population are a part of overall social development, but they show the way that different social groups and subgroups respond differently to health situations. For example cancer rates can improve in one group as they deteriorate in another. If epidemiology only observed individual susceptib-ilities, it would never be able to get at the social causes that produce such unequal risks. Thus it is important for epidemiology to explain the develop-ment of social differences in health within a population, and pass this knowledge on to health promotion.

The ethics and purpose of epidemiology require it to improve health by demonstrating the health relationships of living and social conditions. It is the purpose of medicine, in contrast, to advise patients how to act so as to prevent or cure a disease. But the overall struggle to promote health is not in the end a question of behaviour but of interventions into politics and daily life. Prevention may be, in part, a task of medicine. But health promotion is the challenge for our society.

REFERENCES

Brown, E. R. (1989) *Community Action to Promote Health*. Presentation to the International symposium on community participation and empowerment strategies in health promotion. Symposium of the Zentrum für Interdisziplinäre Forschung der Universität Bielefeld, Bielefeld.

Crawford, R. (1977) 'You are dangerous to your health. The ideology and politics of victim blaming', *Int. J. Health Serv.* 7: 663–80.

Niehoff, J.-U. (1978) 'Risikofaktor-Risikofaktorentheorie-Risikokonzept', *Ztschr. ärztl. Fortb.* 72: 84–9; 72: 145–9.

—— (1979) 'Epidemiologie – methodologische Probleme und Möglichkeiten einer Wissenschaftsdiziplin', *Ztschr. ärztl. Fortb.* 73: 967–74; 73: 1026–9.

Salmon, J. W. (1989) 'Dilemmas in studying social change versus individual change: considerations from political economy', *Health Promotion* 4: 47–52.

9

A THEORETICAL MODEL FOR HEALING PROCESSES

Rediscovering the dynamic nature of health and disease[1]

Marco J. de Vries

We are always changing, always reaching a higher level after each change – yet with the harmony of our life unbroken and unimpaired.

(Winston Churchill)

1 INTRODUCTION

Healing is the art and heart of medicine.[2] To be a healer requires a sensitive heart, an acute mind, and disciplined training (the last requirement is one which tends to be pushed into the background in the 'counter culture'). The concept of healing and the knowledge about the process of healing is so central to medicine, that it is surprising to find that modern medical textbooks devote so few pages to this subject. In some recent textbooks of pathology there is no separate chapter on healing; the subject is discussed under 'inflammation'.

The word 'healing' has common origins with the Greek word *holos*, which has the meaning of 'whole', 'wholeness', 'total', or 'complete'. In this contribution I will use the roots 'whole' and 'heal' in their original sense, meaning wholeness as well as health. Also, I will depart to some extent from the definition of health as given by the World Health Organization: 'Health is a state of *complete* physical, mental and social well-being and not merely the *absence* of disease or infirmity.' Wholeness, as we shall see, is not the same as being happy or living without pain, frustrations or handicaps; wholeness may be achieved in the presence of disease or infirmity.

In the text I define *healing* as a *response of a living system to a break in its integrity* (as in a skin wound). *The process of healing is the recovery of the integrity of an injured system* (Bertalanffy 1968; Laszlo 1973; de Vries 1981). There is only a thin line separating healing, as defined here, and the continuous turn-over and replacement of worn-out elements of the system, such as the cells of the skin or intestinal lining. Also one should keep in mind that through the normal strains of living, breaks and restorations of a system's integrity are occurring all the time on a minor scale. When we move, muscle contractions may cause small ruptures of blood capillaries, which are stopped by thrombocyte aggregation, after which the lining of the vessel is repaired. Vulnerability is the price we pay for being highly evolved.

Healing, and the other processes we are describing here, are a dynamic part of the living system's mechanisms of homoeostasis, self-organization and self-renewal. If the old state of a system cannot be restored, we often see that the system creates new forms and functions in the processes of compensation and transformation. Healing, as an adaptation to a challenging environment, is in line with evolution.

In the following pages I will present a model of healing, discuss conditions which must be fulfilled for satisfactory healing, and delineate some of the attitudes and strategies needed by the healer to support this process. The model and the associated principles and concepts, are intended to be practical. In addition, they are holistic, and may be applied to system levels of a very different character, such as cells, organs, individuals, groups, species, cultures, ecosystems, or even our planet. It appears that this model has the property of homology, as described in systems theory (de Vries 1981). I invite the reader to evaluate the usefulness of the model from his own experience.

2   THE STAGES OF HEALING

Starting from the physical level, pathology clearly demonstrates that the process of healing passes through a number of stages, each of these being indispensable (Spector 1980). Arrest of the process in one of its stages, or skipping a stage, usually results in permanent disability or chronic progressive disease. The stages are described below as discrete steps. However, in practice, stages may overlap and they often may be passed through repeatedly.

2.1 Pain

Pain is not usually thought of as a stage of healing. That it is, is illustrated by the serious consequences of the absence of pain in certain neurological disorders. One example is tabes, a complication of syphilis, in which certain tracts in the spinal cord, carrying pain stimuli to the brain, are affected. In patients with this disorder minor injuries of the joints of the leg, such as all of us experience daily, are not repaired. This may lead to complete disorganization and destruction of the joints of the hip, the knee, or the ankle, in the course of only a few weeks. Here we have a first instance of 'homology', for on the psychological level too it is common that a disruption of the personality can only be healed when pain is experienced or allowed to surface.

Recognizing pain as a stage of healing may help health professionals and patients to accept it and reduce the indiscriminate consumption of pain-killers and tranquillizers.

Pain may be defined as a universal sign indicating disruption, or the impending disruption, of the integrity of a system on any level. It moves the organism to mobilize all its resources for repair, transformation and healing. Our definition of pain applies to other system levels beyond the physical. We may distinguish physical, emotional, mental, existential and relational pain, all indicating a break of integrity on that particular level, in a person, between persons or between a person and his environment. Apart from its intensity, the experience of pain can have many *modalities*. On the physical level the modality of the sensation can be stinging, stabbing, burning, throbbing, gnawing, spasmodic or nauseating. These qualitative distinctions are crucial to medical diagnosis. On the emotional level, sorrow, anger, depression, guilt, shame, and fear can be seen as modalities of pain. On the mental level confusion and disorientation are often experienced as very painful. Existential pain, affecting the integrity of a person as a whole, may manifest itself as despair, existential shame and guilt, and bottomless fear. When the integrity of interpersonal relationships is threatened or disrupted, it may be expressed as pain on all the levels we have discussed, including the body (Metz 1975).

As there are so many functional interconnections between the different parts of the personality, it is clear that these different manifestations of pain may present themselves in various combinations. Physical pain may evoke fear and, reversely, emotional or existential pain may be accompanied by physical pain, such as pain in the region of the heart after, for example, bereavement. Also one has to keep in mind that the intensity of the pain experience may differ considerably between individuals and cultures. Patients with similar X-ray abnormalities of the spine, may present with severe, mild, or complete absence of pain. What causes the great cultural differences in pain experience is largely unknown.

2.2 Provisional recovery of integrity

The hallmark of this vital phase is the temporary bridging of the break. This happens in many different ways, depending on the particular system level involved. In skin wounds (Spector 1980) the haemostasis mechanism forms a blood clot, which later transforms into a crust, thus covering the wound. Wound contraction, a miraculous and mysterious phenomenon, is another mechanism for provisional reduction of the defect. After severe loss of blood or body fluids, a series of complicated mechanisms temporarily restore blood pressure and volume (Spector 1980), providing time for the much slower process of renewal of the various blood elements.

On a higher system level, that of the organism as a whole, a very interesting phenomenon, seen after battlefield and traffic injuries, belongs to this stage. Even after very severe injuries, such as multiple bone fractures, people are able to bring themselves into safety by running away from the premises while not experiencing the expected excruciating pain. Pain only manifests itself after they are safe. This 'analgesia' is probably the effect of endorphine production on the brain's pain centres. In these cases the integrity of the person as a whole is endangered and one's life is temporarily saved, even at the risk of extending or complicating existing injuries.

Denial, the first stage of dying as described by Elisabeth Kubler-Ross, may also be seen as the first stage of healing (Kubler-Ross 1970). After the diagnosis of a serious illness, the fear of annihilation may be so great, that it threatens the integrity of the whole personality. The, usually temporary, denial then provides for the time and space necessary to mobilize all one's resources for reorganizing life and personal transformation. Recognizing denial as a creative stage in the final healing process, will help health professionals to respect its positive aspects, rather than to imposing on their patients the acceptance of reality.

2.3 Destruction and removal

In the course of a healing process, the breaking-up and the elimination of dead, non-viable, or non-functional structures are essential. If not removed, they will delay or even prevent healing. One of the important tasks of a surgeon is to remove dead and suppurating tissue, using the healing knife. From ancient times, *ubi pus, ibi evacua* has been a surgical motto. As with pain, medicine and pathology teach us that something often experienced as negative and to be avoided, in this case destruction, may actually be indispensable for healing.

On the physical level, the breaking-up and removal of dead tissue is the function of the inflammatory reaction and of various types of scavenger cells, attracted by the dead cells. Inflammation is not only a response to infection, as is often erroneously thought, but occurs after any tissue damage, whatever its origin.

It is interesting to point out another homology on higher levels of human system organization, where destruction and removal have a constructive function. In serious life crises, such as after the loss of a spouse or a job, when facing a serious disease or physical handicap, it is often essential for healing one's personal integrity to dissolve and eliminate old thought forms, and value patterns which are no longer meaningful or functional. This process of re-evaluation and letting go, usually involves much pain, confusion, fear, and disorientation.

A very similar phenomenon can be observed in the history of the sciences during a paradigm shift (Kuhn 1970). As we may experience in our time, the dissolution of no longer viable theories, models, and views may take much struggle, time, and patience. As Max Planck, the physicist who originated quantum theory, ironically remarked: 'Science proceeds funeral by funeral.'

This stage of healing can be of importance in interpersonal crises, such as in couples or the family. In family therapy one of the objectives is to make explicit long-standing unrealistic or disfunctional images, expectations, and feelings persons have about each other and their relationship and to help them to express and to let go of such 'dead' constructs. Only thereafter can the relationship be healed.

2.4 Regeneration, reparation and compensation

Regeneration is the replacement of lost elements of a system by elements of the same type. On the physical level, cells, tissues, organs and, in plants and the lower animal species, even complicated structures such as stems and limbs, are regenerated (Sheldrake 1981). Replication of stem cells and their differentiation into specialized cells, such as muscle, bone and vascular tissue, is the basic mechanism in this process. It is important to point out here what seems to be a principle in nature. The higher an organism is on the evolutionary ladder, the lesser are, in general its capacities for regeneration on the biological level. This is especially true for very specialized tissues. A newt can regenerate a whole limb which has been cut off. In the higher animals such powers of regeneration of whole body parts have been lost. In man, cells of highly evolved tissues such as nerve cells and heart muscle fibres, cannot be replaced. This evolutionary phenomenon is closely related to the process of mechanization as described in systems theory (Bertalanffy 1968). The more a system evolves, the more processes on lower system levels become automatic, or deterministic, which involves a loss of the versatility of regulatory mechanisms, such as regeneration.

Although man has little capacity to regenerate his brain tissues, he excels in nature by the regenerative powers of his mind and spirit. One of the many testimonies to this in history and literature is Victor Frankl's book, *Man's Search for Meaning*, in which he describes his survival from the utterly adverse conditions of a German concentration camp (Frankl 1962). Similar accounts have recently been published by patients with serious or crippling diseases (Cousins 1979).

Reparation is the replacement of the lost elements by more primitive elements. After heart attack, lost muscle tissue is replaced by connective tissue which later turns into a scar. Although the scar tissue cannot restore the contractile function of heart muscle, it does provide the support and union needed to repair the defect. Scars, then, pose certain limitations of function on the organizational level involved.

Compensation is the taking over of the function of lost and irreplaceable elements by other elements of the same type. In the case of heart attack, remaining muscle fibres may enlarge and thus increase their contractile power. In infantile paralysis other muscles hypertrophy and, at least partially, take over the work load of the paralysed muscles.

A related mechanism is *reallocation* of lost function to other parts of the system involved. A striking example of the powers of reallocation, even in such tissues as the brain, is provided by some of the hydrocephalic children described by John Lorber (Lorber 1983). In one boy, vision, normally mediated by the back parts of the cerebral hemispheres, in this case absent, appeared to have been taken over by a parietal remnant of the brain.

As we have seen, regeneration may be limited and thus functional limitations may be permanent. However, this only applies to the level of the system involved. A major rule to be observed in the science of healing is that *the integrity of a system level or the system as a whole cannot be judged from the changes on lower levels of organization. Healing may take place in the presence of limitations or permanent loss*. From the microscopic observation of scar tissue at the site of a previous heart attack, one cannot evaluate the function of the heart as a whole. This function may have been quite adequate. Patients whose locomotor system has been seriously crippled by rheumatoid arthritis or infantile paralysis, have been able to live meaningful lives and fully express their life's purposes. Their creativity often even surpasses that of people without limitations on the physical level. One often cited example is the former US president Franklin D. Roosevelt. Similarly, persisting neurotic patterns on certain levels of the personality do not preclude profound personal and spiritual transformation, as we have observed in some cancer patients (a fact which may be hard to accept for some psychotherapists). Just as with the loss of the regeneratory capacity of body parts, it appears that the evolutionary force in nature is more interested in promoting the evolution of the systems as a whole, than that of their parts.

2.5 Restitution, remodelling, and transformation

The final stage of the healing process is probably the most fascinating and at the same time the least understood. Without restitution, remodelling, or transformation, healing will not only remain incomplete, but may eventually lead to serious complications and secondary malfunction. During this stage the new elements formed in the previous stage, are integrated into the organization of the system as a whole. *This means that either the original architecture is restored in accordance with the system's blueprint and programme, or that the system is reorganized as a whole*. The latter event I indicate by the term *transformation*, and it may involve a change of blueprint and programme.

This process closely resembles morphogenesis as occurring in embryonic development. It involves the *alignment* of like elements (as with the newly regenerated epithelial cells at the surface of a wound), the *ordering* of groups of unlike elements (surface epithelium, glands and connective tissue of a wound) and the *junction and interconnection* of the various elements (e.g. the connections between a newly formed vascular bed and the surrounding blood vessels, so that blood circulation in a wound is restored). Just as in morphogenesis, one of the greatest mysteries of this process is the source of information for the programme being expressed in the restoration of function and architecture. It is here that the concept of morphogenetic field (Sheldrake 1981; Goodwin 1984), as I have discussed elsewhere (de Vries 1983), is very helpful in increasing our understanding of the healing process. Even after extensive destruction of forms and structures, the morphogenetic field for the lost parts seems to be somehow preserved and, on the condition that the previous stages are completed, insures completion of healing.

One of the clearest illustrations of remodelling on the physical level, is the undisturbed healing of a bone fracture (Spector 1980). In its first stages the integrity of the broken bone is provisionally restored by a blood clot, into which connective tissue grows. After removal of dead bone fragments, regeneration sets in with the formation of primitive bone tissue, growing out into an irregular mass, called the callus. In the last stage, a process of transformation of the newly formed bone takes place. Scavenger cells eat away superfluous bone. (But how do they know where to stop eating?) Then the bone is ordered into parallel, or concentric layers and these are joined and aligned in precisely the right way to ensure optimal mechanical stability under the forces to which it is, or will be, subjected.

Homologies of this stage can be found on higher system levels. For example, there is evidence that reorganization of one's life plan, life-style and patterns of interpersonal relationships, may improve the chances of healing in ischaemic heart disease and cancer (Lynch 1977; Simonton and Simonton 1975). In children or elderly people who have suffered brain damage, the re-integration into the family structure may be vital to the recovery of the mind–brain interface and physical rehabilitation.

On the other hand, *failure of this stage of healing* after disease, or personal or interpersonal life crises may be reflected in poor physical health. It may be a factor in the manifestation of such common Western illnesses as high blood pressure, cardiac infarction and cancer (Lynch 1977). When reorganization, remodelling and transformation are interfered with, serious pathology usually results. The previous stage, regeneration, may then carry on unchecked, leading to *overgrowth of cells and tissues, and ultimately, total disorganization*. In addition, in the areas affected a *precancerous condition may arise*, which means that the risk that some form of cancer manifests, is increased. For example, cancer of the skin may develop in skin wounds of long duration, in which for some reason, healing is arrested, as in radiation ulcers, inadequately treated burns, and chronic ulcers of the leg due to varicose veins.

A classic example is cirrhosis of the liver, a chronic disease, which in the majority of cases is caused by alcohol abuse. Stages 2 to 4 of the healing process are all active in an attempt to repair the chronically recurring damage to which the liver is subjected. However, especially in long-standing cases, signs of stage 5 are noticeably absent in microscopical sections. Overgrowth of liver cells (regeneration) and connective tissue (reparation) lead to increasing, and ultimately irreversible, disruption of the liver architecture. In the absence of remodelling, newly formed blood vessels remain unconnected to the blood circulation, or form abnormal connections, leading to the circulatory complications of cirrhosis: fatal haemorrhage and ascites. Finally it is known that in about 15 per cent of cases, carcinoma of the liver develops.

Cancer itself is disorganized growth, rather then excessive growth (many cancers have a relatively slow growth rate). We may tentatively conclude that cancer is the result of disturbed (re)organization and transformation. In other words, a failure of the expression of the architectural plan of a tissue. Using cancer as a metaphor, and looking at our planet as a living organism of which we are an integral part, we might say with some justification that Gaia is suffering from cancer. In order for the last stage of healing of our planet to take place, humanity may have to become responsive to the 'morphogenetic' field which has the potential for initiating our transformation.

3 CONDITIONS FACILITATING HEALING

3.1 Rest, protection, and support

On many levels of organization a healing system in its first stages (at least up to the stage where regeneration is well on its way), has a high degree of *vulnerability*. A healing skin wound will easily bleed when it is only covered by a crust or a thin layer of recently regenerated epithelium. At the site of a healing bone fracture the immature bone of the callus is still soft and may be easily bent. Under these circumstances, healing tissues will be damaged by relatively slight insults. Many medical measures with a long historical tradition are based on this insight, such as bed rest, bandages, splints, arm slings and ice bags. These all testify to the acknowledged need for *rest, protection, and support* in the first stages of healing.

These principles also hold for healing on higher system levels. After recent life crises, confrontation with death, interpersonal crises and even in the case of disruptions on the level of organizations, cultures, nations, and political systems, a skilled healer will provide rest, protection and support. At this time, a call for activity on the patient's part, or an appeal to his will-power or responsibility and confrontation by the health professional would be indiscriminate and premature.

If one does not take into account the vulnerability of the system at these first stages, healing may be delayed or arrested. In bone fractures that are insufficiently immobilized, remodelling may be disturbed, or even worse, the ends of the fracture will not join. Large skin wounds, exposed to too many strains in this period, will never recover normal tensile strength. I invite the reader to look at his own experience to find 'homologous' examples on higher system levels. To summarize, on the healer's part this is the time for the qualities of the heart, and not so much those of the sword.

Of the many specific measures relevant in the framework of protection and support, I will mention two that are of special importance, nutrition and touch.

3.2 Nutrition and nourishment

In times of acute crisis, endocrine activity increases, especially of the hypophysis, the adrenal, and the thyroid. This results in greatly increased consumption of the body's stores of carbohydrates, fats and proteins, in order to provide caloric energy. It has recently become clear that patients with serious injuries or shock have a greatly increased calorie requirement (Spector 1980). In the stage of regeneration, when new tissues are formed, the diet should be complemented with extra *proteins*. Very little is known about the specific needs for *vitamins*. Vitamin C is essential for connective tissue formation (reparation), while an adequate supply of *zinc* is important for nucleic acid and protein synthesis (regeneration).

The same applies to aspects of nourishment other than dietary. Much more attention should be given to the nourishing aspects of verbal and non-verbal communication, sound and music, light and colours, literature, and many other needs of the human mind and soul, which might directly facilitate healing. One only has to visit a modern hospital, to fully appreciate the lack of such nourishment.

3.3 Touch

It has recently been found that human touch has profound effects on various physiological functions such as cardiac rate, heart rhythm, blood pressure, and respiration (Lynch 1977). These effects were recorded even in deeply comatose patients. Health professionals, who are always touching patients professionally, should be much more aware of how the way they touch may facilitate, and on occasions interfere, with healing. For example, measures such as massage increase the local blood flow and stimulate nervous activity, both known to be essential ingredients for healing. In addition, pain can be alleviated, so that the use of analgetic and hypnotic drugs and their adverse effects can be reduced, while sleep can exert its regenerative power. Touch has many less tangible benefits.

3.4 Activity, movement, and use

In contrast to the early stages, in the later stages of regeneration and especially in the stage of reorganization, remodelling and transformation, patient's activity is necessary and decisive. Bone fractures are again a good example. Movement, use, and putting weight on the bone stimulate the reconstruction of bone and the realignment of bone layers parallel to the lines of force, ensuring optimal adaptation to the mechanical loads to which it is normally exposed. Lack of activity and disuse will lead to the arrest of healing, progressive porosity and wasting away of bone tissue. In addition, adhesions of muscles, ligaments and joint tissues, which arose during the immobilization period, will become fixed and lead to permanent loss of mobility.

Where permanent limitations are unavoidable, activity will initiate re-organization and transformation, and thus healing on a higher level of the system. This is true for all levels. People with serious limitations of their locomotor system, e.g. as a result of rheumatoid arthritis and neurological disorders, can sometimes demonstrate their astonishing powers of trans-formation. They have changed their body scheme and patterns of movement in such a way that their limitations are only apparent to an experienced observer. It is here that rehabilitation medicine, physiotherapy, and tech-niques which stimulate and improve body awareness, such as the Alexander method, are invaluable. On the psychological level mental activity can heal wounds by developing new behavioural and attitudinal patterns.

3.5 Will and responsibility

Parallel to activity, movement, and use, is the exercise of the will and the responsibility the patient takes for his own healing. Very little is understood about the biological mechanisms by which the will affects physiological functions and healing. It has long been known that those who 'fight for their lives' heal faster and live longer. This is true for minor disorders such as influenza and skin wounds, as well as for major diseases such as cancer and heart attack. On the other hand, one may expect a worse prognosis in patients who have 'given up' or who have lost all meaning in life. In extreme cases, loss of will can cause those without any previous illness to die, for example shortly after a major life crisis, such as the loss of a spouse (Lynch 1977; Engel 1971). It appears that the will is a major morphogenetic field, even on the biological level (Assagioli 1973).

The most elementary level of taking responsibility is the *patient's choice* for the treatment proposed by his physician, whether it is an operation, drugs, radiation, or a diet. It is important to distinguish such a choice from compliance. Indeed, a patient's choice *not* to follow his doctor, may also be an exercise of will, and thus potentially may promote healing, despite any trouble it may cause to the health professional. I emphasize, may be. Refusing treatment may be based on fear of the particular treatment, or on other conscious or unconscious motivations. One has to discriminate carefully, here.

At a higher level of self-responsibility is the difficult task of marshalling the *will to change one's life-style, habits and patterns of behaviour*. For example, to stop smoking or drinking alcohol, changing dietary habits, or the way one responds to life situations and other people.

The drive to give meaning and a sense of purpose to life, even against a background of radical changes of life situation, is the highest level of taking responsibility. The power people develop in doing so, makes them transcend their personal situation, including serious illness, physical limitations, pain, sorrow, and impending death. Paradoxically, on this level one can heal oneself, without being cured on the physical level. In a special sense this is also true of the choice to die, the final stage of healing.

4   THE ATTITUDE AND STRATEGIES OF THE HEALER AND THE HEALING GROUP

Summarizing, the general attitude and strategies of the healer in the early stages should be *supportive*, actively using such measures as are appropriate at that time. In the later stages, the healer's attitude can best be described as *evocative*, in the sense of evoking the will and useful activity on the part of his patient, while he himself takes a back seat, trusting the natural powers of healing, the *vis medicatrix naturae*.

In the new paradigm of science, the concept that one can be a neutral observer and actor in relation to an observation or technical activity, is questionable. We are entering an era in which subject and object, and thus health professional and patient, together form a system. They mutually interact, whether willingly or unwillingly, consciously or unconsciously. Among other things, this means that the way healers think, their views about health and disease, their expectations, and motives, may influence the process of healing in their patients, in both a positive and a negative direction. For example, there is evidence that patients of doctors who believe cancer to be a universally lethal disease, die earlier than the patients of those who believe cancer can be arrested or cured (Simonton and Simonton 1975). (It is important to realize that these beliefs need not be consciously held, in order to have a possible effect). To tell a patient or his family that he has only a few months to live, is a scientific error (survival statistics apply to populations and have little significance for individuals), as well as a serious professional mistake. I have often asked myself to what extent the survival statistics of cancer patients reflect widely held professional beliefs about cancer. How much of patients' compliance is a Western variant of 'Voodoo death' (Cannon 1957)?

There are now good reasons to act from the assumption that the thought forms, the feelings and, as we have seen earlier, the physical activities of a health professional such as touching, may function as 'morphogenetic fields' for his patients. This creates a special responsibility on the part of the healer; a challenge and an opportunity to transform the co-operation between the healer and the patient into a *healing relationship*. In the healing relationship the healer seeks consciously and scientifically to use his or her own person (body, feelings, mind, heart and soul) as a morphogenetic field to support the healing of the patient. Though psychological factors are involved, we have to do here with life processes, based on principles in nature of a much more universal character than merely psychology or psychotherapy. That this is so, is illustrated by the principles of healing discussed earlier, which seem to have validity for many levels of system organization. Under the new paradigm, one of the important tasks for a science of healing would be to investigate scientifically the relatively subtle field effects involved in healing on all levels.

What then are the basic attributes of a true healer? First, a solid knowledge of his field and a disciplined mind. This involves willingness to follow critically recent scientific developments and to let go of ideas and views which no longer fit into the context of developing knowledge or one's own critically evaluated practical experience.

Second, the creation of a healing relationship requires, in addition to a basic attitude of respect and love for life in all its forms, a personal commitment. This commitment is for healers to heal themselves, to become whole persons. As a first step, this means to accept and own their own powers and potentials, as well as their limitations, however difficult this may be. Only then can they be present to their patients. It is this presence which is needed to evoke and support the potential for healing in others. One of the most difficult steps for health professionals, is to accept the pain, the suffering, the fear and the despair of the people whom they are caring for. This often touches them so deeply that it may evoke the illusion of impotence. One of the most meaningful lessons I learned from cancer patients and parents of seriously brain injured children is, that only after accepting totally one's own condition and that of the child, including the limitations that the condition imposes on life and possible life expectancy, can the first steps be made on the path towards healing.

All that has been said before about the attitude and strategies of the individual healer, also applies to the healing group. This is a group of people who have committed themselves to co-operation in the health-care field. This co-operation can take many forms, such as that of a clinic, a hospital, a primary-care health centre, or a family practice unit. For example, the climate of thoughts held by such a group will influence the healing processes of those that visit or are admitted to the particular unit. Such influence will be much more potent than that of an individual. The way the members of a team relate to each other is of equal importance to the climate created. For example, are they concealing their feelings of pain or anger from each other, or are they creating opportunities for those members who are drained or 'burnt out', to be nurtured, comforted and healed?

Just as the fields created by a group may be potent, so are the opportunities for establishing a healing climate. To begin with, this may not be such a big thing as it sounds. Seemingly small details, such as the presence of a nurse in the operating theatre, appointed to the task of talking to or touching the patient, may have profound effects. So may have a joint meditation on the task and the patient prior to the operation. A little reflection reveals that there is so much space here for creativity and research.

5   IN PRAISE OF THE HEALER

Once I was called to our hospital's intensive care unit, where a friend had been admitted with a cardiac arrest. Being with my unconscious friend who was struggling for life, hooked up on to the complicated machinery, I had my usual reaction of anger towards the doctors and nurses, frantically busy with watching dials and displays of electronic equipment, while apparently not seeing my friend. My anger increased when a nurse said to me that talking to and touching my friend was useless, because she was unconscious. And then suddenly my eyes opened. What I saw was an incredible commitment to their job. How else could these young doctors and nurses (some of them still looking like children to me) manage to work day after day and night after night in this atmosphere, being confronted with the pain, agony, and fear, fighting death? They should be playing, enjoying all the beautiful things available to them, like other people of their age do.

Looking back at that episode in the ICU, and many years of teaching medical students and health professionals, it becomes more and more clear to me that those who have chosen to practise medicine in one of its forms, are, in a certain sense, a group of very special people. The best way I can describe my experience of many of those that I have met, is to use the metaphor of the *bodhisattva*. In Buddhism, a bodhisattva is someone who has made the decision to postpone his own liberation in order to alleviate the suffering of others (*Encyclopaedia Britannica*). The most important attribute of the bodhisattva is that he joins in himself the virtues of compassion and wisdom, which historically has also been an attribute of medical practitioners.

A true healer, just like a bodhisattva, has chosen to serve people whose life's integrity has been disrupted by illness and disease. To exercise his service he should be gifted with a great and sensitive heart, which radiates compassion and so becomes, together with his mind, his most valuable instrument. In my experience, many health professionals have little or no consciousness of their gift, nor of the fact that this gift also makes one

141

especially vulnerable. That this is literally and physically true is demonstrated by epidemiological studies which have shown a much higher mortality among physicians from heart disease, cancer, accidents, suicide and drug abuse, as compared to the general population.

Having to face so much suffering and despair in others, as is especially the case for those who daily practise in such modern medical institutions as cancer hospitals, or hospices for dying or seriously debilitated people, may literally overtax a vulnerable heart. It isn't surprising that many health professionals then withdraw in self-defence behind a wall of technicalities, pragmatism and routine. The same may be true for such medical institutions as a whole, explaining their sometimes sterile and inhuman appearance. In addition, much patient suffering cannot even be alleviated by the armoury of present-day medicine, which makes the confrontation even more painful.

On the other hand we have seen that pain and withdrawal are the first stages of healing. Who can be trusted more to heal himself than the wounded healer? To be a healer is seeing through one's own wounds the needs of the other.

REFERENCES

Assagioli, R. (1973) *The Act of Will*, New York: Viking Press.

Bertalanffy, L. von (1968) *General Systems Theory*, New York: George Braziller.

Cannon, W. (1957) 'Voodoo death', *Psychosomatic Medicine* 19: 182–90.

Cousins, N. (1979) *The Anatomy of an Illness as Perceived by the Patient*, London: Norton.

Engel, G. L. (1971) 'Sudden and rapid death during psychological stress', *Ann. Intern. Med.* 74: 771–82.

Frankl, V. (1962) *Man's Search for Meaning*, New York: Simon & Schuster.

Goodwin, B. (1984) 'A relational or field theory of reproduction and its evolutionary consequences', in M. W. Ho and P. T. Saunders (eds), *Beyond Neo-Darwinism*, New York: Academic Press.

Kubler-Ross, E. (1970) *On Death and Dying*, New York: Macmillan.

Kuhn, T. S. (1970), *The Structure of Scientific Revolutions*, Chicago: The University of Chicago Press.

Laszlo, E. (1973) *Introduction to Systems Philosophy*, New York: Harper & Row.

Lorber, J. (1983) 'Is your brain really necessary?', in J. L. Grashuis, J. A. F. Oomen, R. Stoeckart and G. S. A. M. Ven (eds), *The Nature of Reality*, Rotterdam: Studium Generale, Erasmus University, Rotterdam, p. 85.

Lynch, J. J. (1977) *The Broken Heart: the Medical Consequences of Loneliness*, New York: Basic Books.

Metz, W. (1975) *Pijn: een teer punt*, Nijkerk: Callenbach (in Dutch).

Sheldrake, R. (1981) *A New Science of Life*, London: Blond & Briggs.

Simonton, C. O. and Simonton, S. (1975) 'Belief systems and management of the emotional aspects of malignancy', *Journal of Transpersonal Psychology* 7: 29–47.

Spector, W. G. (1980) *An Introduction to General Pathology*, Edinburgh: Churchill Livingstone.

Vries, M. J. de (1981) *The Redemption of the Intangible in Medicine*, London: Psychosynthesis Monographs.

——— (1983) 'The emergence of a new paradigm in medicine: reflections on the nature of reality and medicine', in J. L. Grashuis, J. A. F. Oomen, R. Stoeckart and G. S. A. M. Ven (eds), *The Nature of Reality*, Rotterdam: Studium Generale, Erasmus University, Rotterdam, p. 17.

10

HEALTH WITHIN AN ECOLOGICAL PERSPECTIVE

Lucas Reijnders

1 INTRODUCTION

The emergence of the ecological (or environmental) crisis as one of the main issues of the 1990s is only partly related to the health implications of environmental (or ecological) change. Nevertheless changes in our ecological setting do have important implications for health. The environment or our ecological setting may be defined as the whole of the physical aspects of our surroundings. This definition of the environment includes such aspects as the availability of natural resources, the presence of populations of other species, the availability of space for specified activities and the presence of pollutants. These aspects are subject to change. Some of these changes do currently have substantial negative impacts on health. In the highly polluted 'black' triangle of Southern Silesia, Sachsen (South Eastern Germany) and Northern Bohemia there is suggestive evidence of a substantial negative effect of pollutants on life expectancy (Knook 1991). Third World mega-cities like Mexico, and the Aral Sea area, also clearly illustrate negative health effects of pollution (World Resources Institute 1990; Brown *et al.* 1990). Loss of playgrounds and increased lead levels caused by the increase in motorized traffic have a negative impact on child development in cities. The pollution-induced loss of ozone from the ozone layer increases skin cancer and cataract risks, while reducing the effectiveness of the immune system (World Resources Institute 1990). Increased temperatures, the expected result of increased atmospheric concentrations of greenhouse gases such as carbon dioxide, are expected to increase mortality among the elderly, especially in areas where hot weather is currently uncommon. This is related to increased incidence of heat shock, stroke and heart disease (World Resources Institute 1990). In parts of Africa the loss of such an important natural resource as fertile topsoil owing to erosion creates environmental refugees who are often subject to impaired health (World Resources Institute 1990; Brown *et al.* 1990). Erosion-induced mudflows and inundations are a substantial health hazard in parts of Asia.

To get a clearer understanding of current and expected environmental deterioration, we shall briefly survey three major aspects of the environment. These are the (future) availability of natural resources, our coexistence with living nature, and pollution.

2 NATURAL RESOURCES

Health in a society is partly dependent on the availability of natural resources. There are several kinds of natural resources. One of the most important distinctions in this respect is between renewable and non-renewable resources. In the case of renewable resources, significant amounts are currently being produced. In case of non-renewable resources, there is no current generation of such resources in any significant amount.

Of the renewable natural resources that are currently exploited soil, groundwater and wood are probably the most important ones. These three resources are subject to quantitative and qualitative deterioration (Brown *et al.* 1990; Myers 1987; Pimentel *et al.* 1987; Kaufman and Mallory 1986). The current net worldwide loss of forest is at least 10^7 ha/year (Brown *et al.* 1990; Myers 1987). Moreover the health of forests is threatened in several parts of the world by pollution problems like soil acidification, oxidizing smog and climatic change (World Resources Institute 1990; Brown *et al.* 1990). A substantial number of countries are subject to falling groundwater tables, and pollution of groundwater is an increasing problem, especially in intensively farmed and heavily industrialized areas. The worldwide loss of fertile soil is currently estimated to be at least 20.10^9 tonnes/year (Brown *et al.* 1990). Mainly because of erosion and increased salt content approximately 6.10^6 ha/year of agricultural soils are taken out of production, whereas, owing to erosion, productivity of soils currently used in agriculture is estimated to fall by 15–30 per cent if we maintain current practices over the next 25 years (Pimentel *et al.* 1987; Kaufman and Mallory 1986). As to non-renewable resources the use of geochemically scarce non-renewable substances is especially important. These include fossil carbon fuels, mineral phosphate and the geochemically scarce metals (Reijnders 1989). Many authors have dealt with the future availability of fossil carbon compounds. From their studies, it can be concluded that physical scarcity of natural gas and mineral oil will become a fact of life in the next century if we maintain current rates of use (World Resources Institute 1990; Reijnders 1989). Coal reserves are more abundant, but at current consumption rates they will probably last only a few centuries.

If there is no dramatic change in patterns of use, there will be an end to substantial extraction of a number of geochemically scarce metals in the next centuries. These include copper, lead, zinc, tin, gold, silver, mercury, platinum metals, gallium, vanadium, tungsten, bismuth, zirconium, tantalum, antimony, molybdenum, hafnium, niobium (columbium), yttrium, indium and most of the rare earth metals (Reijnders 1989). Extrapolation of current patterns of phosphate use suggest that presumed resources may last for four to five centuries (Reijnders 1989).

In summary our resource base is being steadily eroded by over-exploitation of renewable resources, pollution and exhaustion of geochemically scarce non-renewable resources.

3 LIVING NATURE

There is a rapid loss of other species and of the land area of virgin or wild nature. The loss of natural species is currently probably in the order of 0.5 per cent per year – roughly 10^4–10^6 times the normal extinction rate (Myers 1987; Reijnders 1989). The loss of wilderness is even more rapid. Several factors contribute to this rapid loss. The main driving force is the expansion of the world's population, and associated food production. Other factors such as logging, pollution, tourism and bad productive practices also contribute. There is no sign of any slowdown in the loss of wild nature and natural species. Negative effects thereof on a micro-scale are significant. This may be illustrated by some examples. A substantial number of medicines, including antibiotics and a number of cytostatics, were 'invented' by living nature. Loss of living nature will restrict expansion of our therapeutic base.

Similarly the base for future food production narrows through the loss of wild nature (Myers 1987). Also loss of forests in mountainous regions increases danger of mud-, rock- and snowslides, as well as loss of topsoil. However, the macro-effects of loss of nature are in the long run probably more important. These macro-effects mainly relate to the important role of living nature in geochemical cycles like the carbon, nitrogen, sulphur and water cycles. These cycles strongly determine our current abiotic environment, including such aspects as atmospheric concentrations of carbon dioxide, oxygen and nitrogen compounds and precipitation (Schneider 1990). By now about 40 per cent of net production of biomass occurs under human control. Characteristics of this production of biomass are as a rule quite different from the characteristics of wild nature. This extends to participation in major geochemical cycles. Often primary production (photosynthesis) is lower in a man-controlled setting than in wild nature. However the emissions of N_2O (nitrous oxide, an important greenhouse gas) and CH_4 are often substantially increased, if compared with virgin nature (CLTM 1990). Evapotranspiration influencing precipitation and groundwater tables is often also substantially changed (World Resources Institute 1990; Brown *et al.* 1990). Further loss of virgin or wild nature will in all probability substantially change our global abiotic environment.

4 POLLUTION

There are different types of pollution. A major distinction is that between flux-type and sink-type pollution (Reijnders 1989). In the case of flux-type pollution, pollutants are moving. Such movements may be fast, e.g. many kilometres a day, as is often the case with river and air pollution. Such movements may also be slow, e.g. centimetres a day in the case of groundwater pollution. Sink-type pollution concerns accumulation of pollutants in a compartment of the environment. Another major distinction relates to the transformation of pollutants. In practice some pollutants are quickly transformed into substances that are less of a problem, 'degraded'. Other pollutants only degrade slowly. Still other pollutants are non-degradable, 'conservative' or transformed into substances that are more of a problem (e.g. mercury into alkyl-mercury). Currently there is growing concern about slow flux- and sink-type pollution by substances that are not readily degradable.

Sink-type pollution includes (Reijnders 1989; Copius Peereboom 1989 and 1991; RIVM 1988 and 1991): deterioration of the ozone layer, gas-induced warming of the atmosphere, soil toxification by a number of conservative (i.e. non-degradable) and poorly degradable substances, acidification of soils and loading of seas with conservative and poorly degradable substances like plastics and increased amounts of CO_2. Sink-type pollution tends to become more severe if a certain level of pollutants added to the environment ('critical load') is exceeded. For instance, it has been calculated that, to prevent a further deterioration of the ozone layer, emissions of fully halogenated chlorofluorocarbons should decrease by about 85 per cent (RIVM 1988 and 1991). Emission cuts in the order of 60–95 per cent have been shown to be necessary to prevent worsening of a number of other examples of sink-type pollution, including cadmium pollution of soils, acidification of forested sandy soils in western Europe and greenhouse warming of the atmosphere by substances like carbon dioxide and nitrous oxide (Reijnders 1989; Copius Peereboom 1989 and 1991; RIVM 1988 and 1991; Stigliani *et al.* 1988).

The most important slow flux-type pollution is probably groundwater pollution. The importance of this type of pollution derives from the importance of groundwater as a resource and ecological determinant, and the long-lasting effect of groundwater pollution, once it occurs. For instance, in a case involving pesticide pollution of groundwater in the Netherlands, it has been calculated that, in order to maintain current standards for drinking water, even if one were to immediately and completely stop the use of the pesticide concerned, purification of groundwater will be necessary for at least one century (Breugelink 1987).

Trends in the different kinds of pollution are variable. In general sink-type and slow flux-type pollution are on the increase virtually everywhere. Fast flux-type pollution is increasing in many third world countries. In industrialized countries there is an increase of some kinds of flux-type pollution, and stability or even a decrease of other kinds. In Eastern Europe sooty sulphur oxide smogs will probably show a decreasing trend. The same holds for pollution of a number of European rivers. However, oxidizing smogs and noise are becoming more of a problem in several industrialized countries.

What is the actual importance of pollution? Though this matter has received much attention there is no full answer to this question. The current situation is still largely characterized by nasty surprises, lack of knowledge and uncertainty. It is clear, however, that induced changes in the concentrations of 'natural substances' like CO_2, phosphate and cadmium strongly contribute to the actual importance of pollution (Reijnders 1989; Copius Peereboom 1989 and 1991; RIVM 1988 and 1991; Stigliani *et al.* 1988). And it is also clear that man-made substances have a major impact (CLTM 1990; RIVM 1988 and 1991; Stigliani *et al.* 1988). There are probably significant contributions of pollution to several major health problems, including increased cancer risk, increases in allergic response and asthma, impaired functioning of the nervous system, and respiratory diseases (Copius Peereboom 1989 and 1991). It is also clear that pollution may lead to the deterioration of the resource base. Examples of such resource-base deterioration include: forest die-back due to acidification of soils and increased airborne ozone, which has effects on wood production and recreation, the negative impact of eutrophication on fishing and the threat to agriculture in low-lying areas due to the rise in sea level associated with the greenhouse effect (Reijnders 1989; Copius Peereboom 1989 and 1991).

5 ENVIRONMENT AND HEALTH IN THE FUTURE

Making reliable predictions about the future is notoriously difficult. This also holds for the future interaction of environment and health. If one looks back on the last thirty years, it appears that several unexpected environmental problems have emerged. These include important determinants of health such as deterioration of the ozone layer, radioactive pollution associated with nuclear energy, the toxicology of polyhalogenated hydrocarbons and destabilization of our abiotic environment owing to loss of wild nature. There is no reason why the next decades should not have similar surprises in store. This makes the future impact of environment on health highly uncertain.

However, extrapolating current trends it seems probable that environmental developments will have a major effect on health. It may well be that in the coming century quantitative and qualitative loss of natural resources will be the most important environmental factor affecting health. In part this loss is directly related to the way resources are exploited. Erosion and salinization of fertile soils, loss of forests causing fuel wood crises and overexploitation of groundwater are typical examples thereof. Pollution, however, will probably also significantly contribute to quantitative and qualitative loss of resources. For instance the expected sea-level rise, associated with global warming, will 'eat away' a substantial amount of low-lying fertile agricultural lands. Pollution of soils will increasingly affect the quality of groundwater and agricultural produce, and thereby the health of those deriving drinking water and food from such soils.

Loss of natural sources of high quality cuts two ways. It will directly affect the health of people dependent on those resources and it will also narrow the resource-base of the curative health-care system. Therefore a major effort to change current trends in environmental change should benefit both preventive and health care.

6   SUSTAINABLE INDUSTRIAL SOCIETY

Surveying current trends, one sees a kind of pincer movement that threatens the viability of industrial society. The fear of such a pincer movement has led to the emergence of the concept of sustainable development. A sustainable development should reflect a 'steady-state' relation between society and the environment. The idea of a sustainable industrial society is not new (Opschoor 1987). In the work of the nineteenth-century political economists like Ricardo and John Stuart Mill, one finds a similar concept, that of the stationary economy. Recently, the concept of sustainable development or steady-state development, has appeared more frequently. The concept has received extra emphasis by being central in the report of the UN World Commission on Environment and Development: *Our Common Future* (Brundtland *et al.* 1987). The definition of a sustainable development is by no means unambiguous. Moreover, with the increasing political importance of the concept, the amount of rhetoric and fuzziness seems to be increasing. The definition of sustainability is strongly dependent on assumptions about the time-span over which a development should remain sustainable and about potential substitution of physical resources by non-physical factors like technology, knowledge and capital.

Starting from the assumption that a sustainable development refers to a time-span of many generations and that there is no substitution of physical resources by non-physical factors in the long run, one may arrive at the following operational definition of sustainability.

- The use of renewable resources should not exceed the current generation of the actual resource. This means for instance that yearly soil erosion should not exceed yearly soil formation and that the harvesting of wood from forests should not exceed forest growth. (Actual sustainable harvests may be lower in the latter case owing to requirements concerning nature conservation.)

- The extraction of geochemically scarce non-renewable resources should be close to zero, unless future generations are compensated by the provision of an equivalent renewable resource. This means for instance that geochemically scarce metals should in principle be fully recycled, and that burning fossil fuels is only acceptable if compensated by for instance generating and storing an extra equivalent amount of energy from wave power, solar radiation or wind energy.
- The extraction of non-renewable resources that are not geochemically scarce, like iron and alumina, is acceptable up to certain extent. However, future generations should be compensated for the extra energy input that becomes necessary when relatively rich and easily accessible resources are exhausted.
- There should be a rapid end to the loss of natural species and wild nature and possibly a roll-back from the current level of man-controlled production of biomass.
- Pollution is unacceptable if it leads to accumulation in one or more environmental compartments, reflected in increasing 'active' concentrations of pollutants. Similarly, slow flux-type pollution, the safety of which is not an established fact, should be considered contrary to sustainability. Exposure to man-made mutagens that may cause mutations in the human germline should be close to zero. Exceptions to these rules may only be made in the case of a guaranteed equivalent compensation to future generations.

Applying this definition of sustainability to real life necessitates a major change in production and consumption. Applying this definition does not mean the end of industrial society. On the contrary, industry is probably essential for sustainability under conditions of high population pressure (Reijnders 1989). However, even with the persistence of industrial production, sustainability will restrict total world population to its present level (Reijnders 1989). This is mainly related to the heavy reliance of sustainable production on renewable resources and the requirement to rapidly end loss of natural species and wild nature.

7 SELF-SUFFICIENCY AND INTERDEPENDENCE

A major element in a transition to a sustainable relation with our environment is a total overhaul of the system of energy production and consumption. This overhaul is necessary to conserve resources and to stop the increase of sink-type pollution problems. The core of the necessary transition is a major improvement of energy efficiency and a shift to energy-supply systems that are powered by 'natural' energy flows such as geothermal heat, solar radiation and wind power. There is much scope for improving energy efficiency. In Western industrialized countries applications of best available (energy-saving) technology would probably cut current fuel use by 60 to 70 per cent, while maintaining the same end-use performance. In Eastern Europe and many Third World countries fuel use may be cut by more than 70 per cent when applying best available technologies, while keeping end-use performance at its current level (Johansson *et al.* 1989; Goldenberg *et al.* 1988; Blok 1984; Krause 1989; Blok 1991). Supplying the energy-efficient consumer with sufficient energy from natural fluxes such as solar radiation is – technically speaking – a real option. In this case there is a premium on bringing supply and demand close together (for instance by equipping homes with solar boilers and facilities for generating photovoltaic energy). Both improvements in energy efficiency and the integration of supply and demand will make energy users more self-sufficient.

While a sustainable relation with the environment may increase self-sufficiency in some respects, it may increase interdependence of actors in other respects. A clear example thereof is the necessity to close cycles, when using non-renewable resources. Closing of the cycle first implies that the use of non-renewable resources should be restricted to the production of substances that can and will be recycled with a high efficiency. Secondly it implies a major expansion of recycling facilities and production based on resources that we currently call wastes. This requires an overhaul of our production system. Currently we have a highly complex and sophisticated industry based on virgin resources. The recycling sector and industry based on non-virgin resources is comparatively small and unsophisticated. In a sustainable relation with the environment it should be roughly the other way round, as far as size is concerned, and all types of production should become sophisticated.

8 PRACTICAL SOLUTIONS

Man's impact on the environment is roughly a function of population size, and of the volume and nature of production and consumption. Limiting this impact therefore may be realized by influencing volume and nature of production and consumption and population size. There are many ways to do so. Some of these ways have been successfully tried by subcultures, other ways are part or are becoming part of mainstream culture. Here we will concentrate on approaches aimed at realizing a sustainable relation with the environment that are appropriate for application by mainstream culture.

8.1 Limiting population size

Stabilization of population is by now a mainstream development in Europe. Here fertility has dropped to a level that if it persists (and apart from immigration) will actually lead to a decrease in population. Some developing countries such as Thailand and Columbia are also doing rather well in limiting population size. Key factors in a robust stabilization of population size seem to be emancipation of women, development of social insurance systems transcending the family level and improved education.

8.2 Lower-input agriculture

In much of agriculture excess resources and pollutants easily leak away. Lower-input agriculture and the substitution of ecologically more acceptable substitutes for damaging inputs is therefore a substantial contribution to a more sustainable agriculture. Much work is being done along these lines both in industrial and developing countries (Proceedings of Conference on Sustainable Agriculture and Integrated Farming Systems 1984; Riedijk 1989; Dahlberg 1986). Interestingly in several cases the above changes have benefited farmers' income or overall agricultural production. Studies in the United States of America have shown that lower-input farming is often more profitable than high-input farming (National Research Council 1989). In Indonesian rice production major reductions in the use of dangerous pesticides have benefited overall food production and reduced overall costs (World Resources Institute 1990). In the Netherlands it was shown that 80–90 per cent cuts in pesticide use and 30 per cent cuts in fertilizer use may actually somewhat improve farmers' income (Proceedings Conference on Sustainable Agriculture and Integrated Farming Systems 1984; Riedijk 1989). One may also note adaptation of farming practices to (re)integration of 'non-productive' species in agricultural production (Dahlberg 1986).

8.3 Prevention of wastes associated with industrial production

Experiences in Western industrial countries with a long tradition of environmental legislation have shown that there is still much scope for prevention of industrial wastes. Good housekeeping, changes in production process, changes in inputs and internal recycling may strongly reduce wastes, while maintaining a given industrial output. Much of these waste prevention options are actually profitable at current prices (NOTA 1991; Freeman 1990; Environmental Protection Agency 1990). Typically a medium-sized industrial company in a Western industrial country may cut production-related wastes by 30 to 60 per cent and make a profit on that. Technically speaking waste reductions of over 80 per cent are often possible by using waste prevention methodology (NOTA 1991; Freeman 1990; Environmental Protection Agency 1990). In Eastern Europe and industries in Third World countries the potential for waste prevention may even be larger.

8.4 Improving energy efficiency

As was pointed out above the technical potential for improvement of energy efficiency is still very large. At current fuel prices not all of these improvements in energy efficiency are profitable. However, in Eastern Europe cuts in fuel use of up to 40–50 per cent may be profitable, whereas in Western countries an overall 30 per cent improvement in energy efficiency will bring a profit on investment (CLTM 1990; Johansson *et al.* 1989; Goldenberg *et al.* 1988; Blok 1984; Krause 1989; Blok 1991).

8.5 Switching away from non-renewable primary energy

Technically a major switch away from non-renewable fuels (fossil carbon, uranium) is feasible. Geothermal energy, wind power and solar energy may take the place of much of current fuel use. The main impediment to this switch is price. However, photovoltaic energy (solar electricity) is rapidly becoming profitable as a source of primary energy in applications that are remote from existing electricity grids. Solar warming of tapwater is already profitable in many 'sunny' countries. Also if costs of pollution were to be added to the price of electricity from coal-fired power stations, wind power would be profitable in substantial parts of north-western Europe. With rising interest in 'ecotaxation' and application of the 'polluter pays' principle, switching away from non-renewable primary energy could be profitable.

8.6 Recycling (Amro Bank 1989; van Weenen 1990)

Recycling of 'wastes' is a necessary ingredient of a sustainable relation with the environment. Interestingly in this respect Third World countries are often way ahead of industrialized societies. Technically speaking there are abundant possibilities in industrial countries. Barriers to such activity mainly come from faulty prices and a lack of design-for-recycling tradition. Including pollution-related costs in product and production prices would probably strongly favour a recycling society. Organic wastes may be generally recycled into fertilizers and/or composts – the latter being useful for improving soil structure. For other wastes, product reuse and material reuse are viable options – especially if design-for-recycling of products and production becomes a way of life. When going for material reuse high-grade applications should be preferred. When the material concerned deteriorates while in use cascade concepts are useful, according to which recycled materials slowly go from high-grade to relatively low-grade applications.

8.7 Green consumerism

One of the striking aspects of the recent past is the widespread emergence of green consumerism, environmentally conscious buying behaviour. Green consumerism has become a market force to be reckoned with. Not only in the industrialized states, but also in countries like Indonesia, Malaysia, the Philippines and Brazil. By now many consumer organizations regularly focus on environmental aspects of products and production. Environmentally oriented product boycotts have become a fact of life in countries as different as Indonesia and Sweden. Many green consumer guides have been published, with topics varying from nappies to building materials, from paper to travel and from motorcars to food (Porrit 1991).

REFERENCES

Amro Bank (1989) *Recycling, een zorg minder voor het milieu*, Amsterdam.
Blok, K. (1984) *Onbeperkt houdbaar*, Utrecht: Stichting Natuur en Milieu.
—— (1991) *On the Reduction of Carbon Dioxide Emissions*, Utrecht: Rijksuniversiteit Utrecht.
Breugelink, G. P. (1987) 'Grondontsmetlingsmiddelen in het grondwater, voorlopers von een ernstige verontreiniging', H_2O 20, pp. 522–6.
Brown, L. R., Flavin, C. and Postel, S. (1990) *State of the World*, Washington, DC: World Watch Institute.
Brundtland, G. H., Khalid, H. and MacNeil, J. (World Commission on Environment and Development) (1987) *Our Common Future*, Oxford: Oxford University Press.
CLTM (1990) *Het milieu*, Zeist: Kerkebosch.
Copius Peereboom, J. W., and Reijnders, L. (1989 and 1991) *Hoe gevaarlijk zijn milieugevaarlijke stoffen* 1 (1989) and 2 (1991), Meppel: Boom.
Dahlberg, K. A. (ed.) (1986) *New Directions for Agriculture and Agricultural Research*, New Jersey: Rowman & Allenhead.
Environmental Protection Agency (1990) *Pollution Prevention Research Plan*, Report to Congress, Washington, DC: US Government Printing Office.
Freeman, H. M. (1990) *Hazardous Waste Minimization*, New York: McGraw-Hill Publishing Company.
Goldenberg, J., Johansson, T. B., Reddy, A. K. N. and Williams, R. H. (1988) *Energy for a Sustainable Development*, Washington, DC: World Resources Institute.
Johansson, T. B., Bodlund, B. and Williams, R. H. (eds) (1989) *Electricity*, Lund: Lund University Press.
Kaufman, L. and Mallory, K. (eds) (1986) *The Last Extinction*, Cambridge, Mass.: MIT Press.
Knook, H. D. (1991) *Internationale Spectator* (November).
Krause, F. (1989) *Energy Policy in the Greenhouse*, El Corrito: IPSEP.
Myers, N. (1987) *Nature Conservation at Global Level: The Scientific Issues*, Utrecht.
National Research Council (1989) *Alternative Agriculture*, Washington, DC: National Academy Press.
NOTA (1991) *Kiezen voor preventie is winnen*, Den Haag: Staatsuitgeverij.

Opschoor, J. B. (1987) *Duurzaamheid en verandering*, Amsterdam: Vrije Universiteit, Amsterdam.

Pimentel, D. and Pimantel, M. (1987) *Bio Science* 37: 277–83.

Porrit, J. (1991) *Save the Earth*, London: Dorling Kindersley.

Proceedings of Conference on Sustainable Agriculture and Integrated Farming Systems (1984) Michigan.

Reijnders, L. (1989) *Naar een nieuwe ijzertijd*, Amsterdam: Van Gennep.

Riedijk, W. (ed.) (1989) *Appropriate Technology in Industrialized Countries*, Delft: Delft University Press.

RIVM (1988 and 1991) *Zorgen voor morgen* 1 (1988) and 2 (1991), Bilthoven.

Schneider, S. (1990) *Environment* (May) pp. 5–32.

Stigliani, W. M., Berendt, H. and Brouwer, F. (1988) *Future Environments for Europe*, Luxembourg: IIASA.

Weenen, H. van (1990) *Waste Prevention Theory and Practice*, Dissertation, Delft: Technische Universiteit.

World Resources Institute (1990) *World Resources 1990–1991*, Oxford: Oxford University Press.

Part IV
METHODOLOGY

11

WHOSE HEALTH? RE-EXAMINING PUBLIC HEALTH POLICY

Jack Warren Salmon

1 INTRODUCTION

The purpose of public health policy should be to prevent disease, to reduce death and disability, and to promote health on a community-wide basis. Public health practice should be grounded in science, in particular knowledge of the multiple aetiologies of various diseases and methods of transmission, as well as methodologies for prevention and for the operation of community health programmes. In essence, public health policy ought to encompass a vast range of concerns for a population's health and social well-being. Its benefits should be improved health status within the national population.

Yet, in practice, public health policy is often more limited in scope than what is required, owing to the political environment in which it is formulated and implemented (Litman and Robbins 1990). In all societies, there is an inevitable compromise in addressing needs for public health and health-care services, owing to the limited resources allocated. Economic constraints upon the health sector are imposed by those in power and by those paying for it. Moreover, constraints emerge that are internal to the health sector. The interests of professionals, health-care institutions, and corporate entities within the 'medical industrial complex' also compromise plans to ameliorate and eradicate disease within certain populations (Salmon 1990). Thus, tensions in public health policy work against its effectiveness.

This chapter examines the notion of public health policy in light of the need for evolving a new 'science of health' (Lafaille 1993). It is intended to provide an informative, albeit brief, review to aid in understanding the role of the policy-making process in health and disease and to demonstrate that issues of public health policy must be included in our rethinking the nature of health.

2 THE LIMITS OF MODERN MEDICINE AMIDST A NEW HEALTH AWARENESS

Discussions on the definition of health centre on the question, 'Where is health created?' These discussions emanate from a growing recognition of the limits of Western scientific medicine and how it is implemented within conventional health-care systems in the contemporary world.

Western scientific medicine emerged from the allopathic medical tradition in the late nineteenth century after the discovery of the germ theory of disease. It brought laboratory and other methods of sophisticated enquiry to aid the understanding and treatment of infectious diseases (Salmon and Berliner 1980). Its dominance over all previous systems of medicine (e.g. homoeopathy, ayurvedic, Arabic, Chinese traditional, and many other age-old forms) became assured through its theoretical exposition, its clinical elaboration and its technological advancement. Scientific medicine has chiefly been focused on the disordered biological state of the individual (Engel 1977).

Western superiority in economic, technological and military spheres per-petrated the assumption that scientific medicine is likewise far superior to all predecessors and competitors. The referent, 'scientific', implied having reached a zenith, assuming prior systems of medicine are based upon lesser foundations (Salmon 1984a).

There is a new critique of the biomedical paradigm. Western medicine has come under attack for its mechanistic and reductionistic approach to human biology (Capra 1982). Besides being viewed as costly and ineffective against many chronic degenerative conditions, this somatic, disease-oriented medi-cine, which historically fashioned all modern health-care institutions, now seems fundamentally flawed (Argument-Sonderband 1983; Illich 1975; McKeown 1976; McKinlay and McKinlay 1977; Powles 1973).

Greater popular attention is being given today to a more ecological and holistic approach to understanding human health and illness (Carlson 1975; Hastings, Fadiman and Gordon 1980; Salmon and Göpel 1990). Many people are reconsidering their health as an experience of well-being, result-ing from a dynamic balance that involves the physical, social, and spiritual aspects of the whole person interacting within our natural and social en-vironments (Blattner 1981; Flynn 1980a, 1980b). This growing public aware-ness about health contradicts an ever-increasing corporatization of health services based on profit motives (mainly in the United States), and conflicts with marketplace mechanisms injected into many countries' health policies across the 1990s (Salmon 1989a, 1990).

In the United States, scientific medicine triumphed over competing med-ical sects with the aid of substantial Rockefeller and Carnegie philanthropy after the turn of this century (Berliner 1985). As a social institution, scientific medicine embraced the dominant ideology of capitalism, and created a set of values and conceptions that have generally been compatible with the prevailing economic order and its social relations (Navarro 1976; Waitzkin 1978).

The focus of medical resources has remained on the individual's medical diagnosis, functional ability, and productive capacity. Over time, the role of society and the environment in disease causation and treatment became downplayed (Totman 1979). As a result, social conditions, public policies, and the nuances of modern urban life have not been considered relevant by conventional medical practitioners (Treacher and Wright 1983).

It is therefore unlikely that the delivery system that developed under scientific medicine is able to foster ideas and responses related to the social, occupational and environmental origins of disease. Nor would it encourage collective responsibility for health; population-based planning; democratization of health knowledge and skills; local community control of health care facilities; social transformations for health improvements; or even effective health promotion strategies targeted for the most ill and disadvantaged (Tannen 1990).

Thus, it is timely to consider how a new science of health may advance beyond incomplete biomedical conceptions, which neglect larger social aspects of disease causation and healthy living, as well as the deeper dimensions to human existence (Salmon 1984b).

3 THE NEW PUBLIC HEALTH MOVEMENT

Because health is created through a series of political, economic, social, cultural and biological processes, more recent attempts to support broader strategies for health improvements within the general population have advanced through the World Health Organization's (WHO) 'Health for All by the Year 2000' campaign. WHO has taken up ideological leadership to incorporate a new definition of health and to stimulate creation of a new public health movement across national boundaries. Within the WHO Alma Ata Declaration (1978), the Ottawa Charter for Health Promotion (1986), and the Adelaide Recommendations on Healthy Public Policy (1988), there is a different sense of health. This difference is based upon the knowledge that it should be a fundamental resource to the individual, to the community, and to the society as a whole.

The advancement of health must be supported through 'sound investments into conditions of living that create, maintain, and protect health. Public health is ecological in perspective, multisectoral in scope, and collaborative in strategy. It aims to improve the health of communities through an organized effort. The themes are: advocacy for healthy public policies and supportive environments; enabling communities and individuals to achieve full health potential; and mediating between different interests in society for the pursuit of health' (Kickbusch 1989: 25).

In the United States, the scope of health promotion activities is much more limited than in Western Europe and elsewhere (Salmon 1991, 1989b; WHO Regional Office for Europe 1985). US conservatives have constricted its definition mainly to an emphasis on individual responsibility for health (Knowles 1977). As a consequence, most of the programmes merely seek to change personal life-styles so as to eliminate 'risk behaviours' for disease control.

The US Surgeon General's report, *Healthy People 2000* (1991), and before that, *Healthy People: The Surgeon General's Report on Health Promotion and Disease Prevention* (1979) and *Promoting Health / Preventing Disease: Objectives for the Nation* (1980) promulgated the ideological tone set by the federal government for the reduction of mortality rates and disabling conditions. Their objectives were primarily aimed at strategies that address controllable risk in the individual, though more comprehensive suggestions for policy change are included.

In the light of substantial Reagan and Bush administration cuts to public health programmes, US policies differ from those of several European nations which embrace the WHO Ottawa Charter to reduce environmental risks and to narrow the inequities in health. For the US, attempts to change personal behaviours have indeed led to improvements in health, particularly among the more educated, highly motivated and better informed (National Center for Health Statistics 1989). Yet, vast disparities in health remain, and for many disease indicators, they have become worse for the less fortunate who have limited means, unsatisfactory social support, and diminishing health-care resources addressed to their needs.

The US official document *Promoting Health / Preventing Disease: Objectives for the Nation* (1980) set 226 measurable objectives, one-quarter of which were not achieved in the last decade. The easier targets were those in which secondary preventive interventions (i.e. screening for high blood pressure, cervical smears, mammography, etc.) were encouraged within the present delivery system, but not without impact on medical costs. Such policy support for early detection and intervention can achieve substantial improvements with proper outreach to at-risk populations. The priority areas that showed progress were high blood pressure control, immunizations, control of infectious diseases, smoking, and alcohol and drug use. Nevertheless, these remain significant problems within the lower class in the 1990s. Areas in which objectives were not at all met include pregnancy and infant health, nutrition, family planning, sexually-transmitted diseases, and occupational safety and health. US Public Health Service programmes addressing these specific problems faced large budget cuts under the Reagan administration. The spectre of AIDS across the 1980s and 1990s complicates all other health-promotion efforts.

Mechanisms for health protection remain problematic in the US, and elsewhere. These are the regulatory and environmental actions directed at larger population groups to protect them from unintentional injuries; occupational safety and health assaults; food, drug and consumer product dangers; and air and water impurities. There are continued limits, particularly during periods of economic recession, in the capability of public health policy to more directly alter patterns of morbidity and mortality through such health protection means.

In the midst of a fervour for a new public health movement among the nations of the world, there are calls for a healthy public policy designed for entire populations (Milio 1986). At a minimum, there must be assurances for decent standards of living, adequate nutrition, safe working and living conditions, proper housing, education, the removal of ecological hazards, and peace; these must become fundamental policy supports to health promotion and protection strategies (Terris 1991). Moreover, as Rosenbrock (1989: 37) has advocated, support for consumer health interest groups in communities and at workplaces has historically been found to create 'institution-framing conditions for preventive medicine and health promotion'. In cities where such actions have become more commonplace (partly encouraged by the WHO 'Health for All' and 'Healthy Cities' programmes, as well as community and labour struggles), there is greater potential for dialogues which investigate the larger social realm and its interrelatedness to health advances within population groups.

When health and social problems become defined in their proper social context, groups of people tend to identify, and thus challenge, existing power relations – either those imposed by professionals and health bureaucrats, or those inherent in larger social structures. The contextual rethinking of the nature of health, and disease aetiologies, helps people to address broader determinants of their quality of life (Brown 1991). This would include health hazards in their more immediate environment, for example those arising from poor food quality. It also holds the potential to minimize victim-blaming tendencies, or at least to understand the social causes of many health-limiting attitudes, behaviours, and conditions (Minkler 1992).

Lafaille (1993) has noted that a science of health has to be inclusive and diverse to conceptually address the interrelatedness of individuals in their communities, societies, and ecological systems. He reminds us that it is ethical to promote free choice, and commitment to care, in a way which implies political, as well as personal, responsibility.

In societies where a significant portion of the population lives in poverty, lacking economic opportunities and political power, it is evident that public health policy will face failure. It should be noted that poverty increased in most nations of the world across the 1980s, with an exponential growth expected across this decade. In the United States, poverty has been found to be much longer lasting. Without appropriate tax policies and social welfare interventions to lift families out of poverty (as one might see in the Netherlands, Sweden, etc.), there can be no likelihood of short term amelioration for the estimated 40 to 50 million Americans near or in poverty. For example, at the turn of the 1990s, one in five US children lives in poverty, a total of over 13 million poor children. Continued social assaults upon this most vulnerable group indicates a terribly unsound investment in future economic and social development.

The poor health status of minority and lower income populations is documented in US health statistics, as well as in the utilization of publicly provided health care resources (National Center for Health Statistics 1979). Higher infant mortality, lower birth weights, developmental disabilities and growth retardation, sexually transmitted diseases, lead poisoning, measles epidemics and other preventable and chronic childhood diseases, mark the fate of this younger segment of the American population. Greater morbidity across all diseases, both infectious and chronic, and shorter life expectancies similarly affect the adult population of people in poverty.

How a society designs its public health policy to treat its children, care for and sustain its aged, and meet the needs of its sick and disabled reveals that society's commitment to health promotion, and, more importantly, its level of compassion for its citizens. The allocation of resources through public health policy is always resolved in the political arena through a political process. Across the 1980s and 1990s, conservative governments in the US and certain European nations introduced a major shift in social and economic thought within their respective countries. Several now face an altered relationship of government (especially, in health and social welfare services) to the individual, family, and community. This new relationship particularly affects those outside, or on the margin of, the power structure. While Europe and the United States present societal models vastly different from the Third World, a similarity rings clear; inequities in health status primarily reflect the distribution of political power and economic resources (Navarro 1976, 1986). Therefore, a primary objective for future research should be to design investigations which ultimately aid new forms of social and political action to redress growing health inequities.

Looking towards the close of this twentieth century, economic contraction in almost all nations gives reason to be pessimistic about the chances of humanity achieving significant improvement in health status and its necessary correlate, greater social justice. In fact, the notion of social progress itself as eliminating social and ecological assaults on health, particularly in the Southern Hemisphere, has become suspect. A growing cynicism is prevalent, even in the 'advanced' societies in North America and Western Europe, where historical promises by governmental leaders have often turned up empty. People are finding it difficult to imagine greater equity in social structure. With the collapse of the Soviet Union and the monumental changes occurring throughout Eastern Europe, threatening migrations, unemployment and inflation may constrain even Western European nations in their striving for 'Health for All'. A profound scepticism exists within the minds of many intellectuals, as well as the mass of the middle and working classes, concerning how much can really be achieved under economic contraction and environmental degradation. New demands for consumption – whether among previous have-nots in the world, or arising from the unabated appetites of past beneficiaries – portends limited social investment for health protection and for promotion of general social welfare.

How will each nation fare under 'free market capitalism', with its accompanying imposed austerity policies and harsh discipline of its labour force? The entirety of the World Health Organization's 'Health for All by the Year 2000' depends upon advancing democracy, or at least providing a more decent alternative to the authoritarian and repressive governments seen in many poor and ruined societies. Again, wider political economic, social, and cultural conditions in each nation, and in the unifying capitalist world, pose deep challenges to groups seeking to create health for all.

For effective public health policy, the social development must be regarded as integral to personal development. A science of health must focus on contextual concerns as a part of the reality of health creation in the individual and in populations.

4 COMPREHENDING THE 'ENVIRONMENT'

In most professional practice within conventional medical settings, the 'environment', when regarded at all, is usually seen as the immediate circumstances of the individual – the composition of people, place, and object surroundings. However, non-physician or alternative practitioners tend to conceive it as an interactional field in which the individual accommodates or adjusts to the prevailing ideologies, social expectations, public policies, and general conditions. This definition of environment may appear abstract and impractical for clinical practice. Therefore it is inadequately addressed in modern medicine and public health policy, and this discourages an explanation for, let alone action on behalf of, persons who refuse or reject accommodation to their existing conditions.

Academic social science disciplines and health professions are beginning to enlarge their research agendas to consider more clearly how the family, workplace, social networks, community, and larger social structures interact. The push to investigate positive health holds promise. For example research on psychosocial influences is now assisting people in managing stress and staying well (Antonovsky 1987). Lessons from the study of psychoneuroimmunology may clarify some external effects on the individual's health and illness. These approaches may uncover contextual origins of contemporary morbidity and mortality, which are thought to be obscured by the focus on the individual (Eyer 1984). This reversing of the lens through which health is examined will support ways to enhance the healing process in individuals and groups, or, at least, will help identify barriers which structurally impede healing.

The causes and the reasons for the spread of disease in specific groups will only be illuminated through more elaborate examinations of their social context (McKinlay 1984). This must reach beyond empiricist traditions dominant in the social and medical sciences, and it requires accompanying qualitative methodologies to identify causative, associated and contributive linkages between the context of people's lives and disease patterns, and their salutogenesis (Ratcliffe 1993). This is not easily done within the methodological constraints of conventional scientific enquiry. Here developments in different scientific fields may lend support to directions for a new science of health.

5 NEW DIRECTIONS FOR A SCIENCE OF HEALTH

Health must be defined and operationalized to embody the totality of the social and ecological landscape, and not merely be confined to the individual's immediate milieu (Turner 1986). Larger power relations, social class relationships, political forces, economic interests, public policies, and ideologies influence people in their worlds. Each is dynamically linked to whole sets of personal behaviours and, more importantly, to corporate and government behaviours that may interfere with health and lead to disease states among groups of people. Unfortunately, the political orientation arising from the dominant ideology has meant that Western scientific medicine has not adequately investigated such linkages. Instead it has merely documented the specific distribution of health in given populations. Movement toward a science of health requires that such external dimensions to the health of individuals and populations be prominently placed in future studies.

Substantive scholarship needs to be employed to clarify ambiguities surrounding positive health (Antonovsky 1987). This should lend direct assistance to those involved in the promotion of healthy public policies (Milio 1986, 1989). More importantly, studies are needed for popular health movements to aid their actions towards altering public health policy, the environment, social and economic conditions, and power relations and structures. Ultimately, a science of health must support, and gain support from, this awareness.

REFERENCES

Antonovsky, A. (1987) *Unraveling the Mystery of Health*, San Francisco: Jossy-Bass.

Argument-Sonderband (eds) (1983) *Alternative Medizin*, Argument-Sonderband AS 77, Berlin: Argument Verlag.

Berliner, H. S. (1985) *A System of Scientific Medicine*, New York: Methuen.

Berliner, H. S. and Salmon, J. W. (1980) 'The holistic alternative to scientific medicine: history and analysis', *International Journal of Health Services* 10 (1): 133–48.

Blattner, B. (1981) *Holistic Nursing*, Englewood Cliffs, NJ: Prentice-Hall, Inc.

Brown, R. E. (1991) 'Community action for health promotion: a strategy to empower individuals and communities', *International Journal of Health Services* 21(3): 441–56.

Capra, F. (1982) *The Turning Point: Science, Society, and the Rising Culture*, New York: Bantam Books.

Carlson, R. J. (1975) *The End of Medicine*, New York: John Wiley and Sons.

Engel, G. (1977) 'The need for a new medical model: a challenge for biomedicine', *Science* 196: 129–36.

Eyer, J. (1984) 'Capitalism, health and illness', in J. B. McKinlay (ed.), *Issues in the Political Economy of Medical Care*, New York: Methuen/London: Tavistock Publications.

Flynn, P. A. R. (1980a) *Holistic Health: The Art and Science of Care*, Bowie, Md: Robert J. Brady Co.

—— (1980b) *The Healing Continuum: Journeys in the Philosophy of Holistic Health*, Bowie, Md: Robert J. Brady Co.

Hastings, A. C., Fadiman, J. and Gordon, J. W. (1980) *Health for the Whole Person*, Boulder, Colo.: Westview Press.

Illich, I. (1975) *Medical Nemesis*, New York: Pantheon.

Kickbusch, I. (1989) 'Prospects for action on health promotion', *Proceedings of "The Challenge of Health: The New Role of Sickness Funds and Health Insurance Schemes"*, International AOK/WHO Conference, Hamburg, FRG, June.

Knowles, J. H. (1977) 'The responsibility of the individual', *Daedalus* 106: 57–80.

Lafaille, R. (1993) 'Towards the foundation of a new science of health: possibilities, challenges, and pitfalls', in R. Lafaille and S. Fulder (eds), *Towards a New Science of Health*, London: Routledge.

Litman, T. J. and Robins, L. S. (1991) *Health Politics and Policy*, Minneapolis, Minn.: Delmar Publishers, Inc.

McKeown, T. (1976) 'A historical appraisal of the medical task', in G. McLachlan and T. McKeown (eds), *Medical History and Medical Care: A Symposium of Perspectives*, New York: Oxford University Press.

McKinlay, J. B. (1984) *Issues in the Political Economy of Medical Care*, New York: Methuen.

McKinlay, J. B. and McKinlay, S. (1977) 'The questionable contribution of medical measures to the decline of mortality in the United States in the twentieth century', *Milbank Memorial Fund Quarterly / Health and Society* 55: 405–28.

Milio, N. (1986) *Promoting Health Through Public Policy*, Ottawa: Canadian Public Health Association.

—— (1989) 'Conceptual framework for a healthy public policy', in J. W. Salmon and E. Göpel (eds), *Community Participation and Empowerment Strategies in Health Promotion*, Bielefeld, FRG: Centre for Interdisciplinary Research (ZIF).

Minkler, M. (1992) 'Community organizing among the elderly poor in the United States: a case study', *International Journal of Health Services* 22(2): 303–16.

National Center for Health Statistics (1989) *Health, United States, 1989* (PHS) 90–1232, Washington, DC: USNCHS.

Navarro, V. (1976) *Medicine Under Capitalism*, New York: Prodist.

—— (1986) *Crisis Health and Medicine*, New York: Routledge & Kegan Paul.

Powles, J. (1973) 'On the limitations of modern medicine', in *Science, Medicine and Man* 1: 1–30.

Public Health Service (1979) *Healthy People: The Surgeon General's Report on Health Promotion and Disease Prevention*, US Government Printing Office, Washington, DC.

Public Health Service (1991) *Healthy People 2000: National Health Promotion and Disease Prevention Objectives*, US Government Printing Office, Washington, DC.

—— (1980) *Promoting Health / Preventing Disease: Objectives for the Nation*, Washington, DC: US Government Printing Office.

Ratcliffe, J. (1993) 'Integrative, transdisciplinary research methodology: principles and application strategies', in R. Lafaille and S. Fulder (eds), *Towards a New Science of Health*, London: Routledge.

Rosenbrock, R. (1989) 'Framing conditions for health promotion and preventive medicine', *Proceedings of "The Challenge of Health: The New Role of Sickness Funds and Health Insurance Schemes"*, International AOK/WHO Conference, Hamburg, FRG, June.

Salmon, J. W. (ed.) (1984a) *Alternative Medicines: Popular and policy perspectives*, New York: Methuen Inc. and London: Tavistock Publications.

—— (1984b) Introduction, Salmon, J. W. (ed.) *Alternative Medicines: Popular and Policy Perspective*, New York: Methuen Inc. and London: Tavistock Publications.

—— (1989a) 'Possibilities for and constraints upon prevention under the health care market system in the United States', *Proceedings of "The Challenge of Health: The New Role of Sickness Funds and Health Insurance Schemes"*, International AOK/WHO Conference, Hamburg, FRG, June.

—— (1989b) 'Dilemmas in studying social change vs. individual change in health promotion: considerations from political economy', *Health Promotion: An International Journal*, 4(1): 43–9.

—— (ed.) (1990) *The Corporate Transformation of Health Care, Part I: Issues and Directions*, Amityville, NY: Baywood Publishing Company.

—— (1991) Introduction to the 'Special section on health promotion strategies', *International Journal of Health Services* 21(3): 417–21.

—— (ed.) (1993) *The Corporate Transformation of Health Care, Part II: Perspectives and Implications*, Amityville, NY: Baywood Publishing Company.

Salmon, J. W. and Berliner, H. S. (1980) 'Health policy implications from the holistic health movement', *Journal of Health Politics, Policy and Law* 5(3): 536–53.

Salmon, J. W. and Göpel, E. (eds) (1990) *Symposium Papers of the "International Symposium on Community Participation and Empowerment Strategies in Health Promotion"*, ZIF Centre for Interdisciplinary Studies, Bielefeld, FRG.

Tannen, L. (1990) 'Health planning as regulation', in J. W. Salmon (ed.), *The Corporate Transformation of Health Care*, Amityville, NY.: Baywood Publishing Co.

Terris, M. (1991) 'Public health policy for the 1990s', *Annual Review of Public Health* 11: 87–98.

The Adelaide Recommendations on Healthy Public Policy (1988) *Health Promotion: An International Journal* 3(2): 183–6.

The Ottawa Charter for Health Promotion (1986) *Health Promotion: An International Journal* 1(4): iii–v.

Totman, R. (1979) *Social Causes of Illness*, New York: Pantheon.

Treacher, A. and Wright, P. (eds) (1983) *The Social Construction of Medicine*, Edinburgh: University of Edinburgh Press.

Turner, J. (1986) World Health Organization Charter for Health Promotion, in *Lancet*, 13 December, p. 1407.

Waitzkin, H. (1978) 'A marxist view of medical care', *Annals of Internal Medicine* 89: 264–78.

World Health Organization (1978) *Declaration of Alma Ata on Primary Health Care*, Geneva: World Health Organization.

World Health Organization Regional Office for Europe (1985) *Targets for Health for All by the Year 2000*, Copenhagen, Denmark.

12

INTEGRATIVE, TRANSDISCIPLINARY RESEARCH METHODOLOGY
Principles and application strategies
John W. Ratcliffe

1 INTRODUCTION

There is a virtual consensus among the world community of nations today that we face a set of international crises of unprecedented magnitude: poverty, famine, and increasing population pressures; ecological destruction, resource depletion, and increasing environmental pollution; materialism, militarism, the spread of arms, and violent revolution (cf. Brown 1984–8). These interrelated problems threaten both our individual health and collective survival. Hence there is urgent need for an in-depth understanding of the complex interrelationships between these problems, for any actions taken will be limited by such an understanding. Each problem has already been addressed one way or another in one forum or another with very meagre results. There is little evidence that this system of problems has been considered in as integrated and coherent a way as it demands, apparently because the traditional approach to scientific research and problem-solving constrains our ability to conceptualize problems of this magnitude. This point is often not well understood, and deserves elaboration.

2 PARADIGMS AND THE PRACTICE OF SCIENCE

To engage in the process of conceptualizing or modelling a problem – that is, to represent it symbolically – is necessary for enquiry into its nature (Mitroff and Sagasti 1973; Ratcliffe 1983). Hence, what is known about any given problem depends both on how the problem was conceptualized and on how the information was obtained. That is, *different conceptualizations of the same problem will produce different information and, therefore, different solutions*. And how a problem will be conceptualized is determined by the worldview or paradigm[1] of the researcher. Currently, there are two paradigms contending for dominance in our Western world – the traditional paradigm and a new, emergent paradigm. The salient differences between each of these paradigms are briefly detailed in the following two sections.

2.1 The analytic paradigm

The paradigm that currently dominates the thinking and guides the practice of the larger scientific community today is known as the analytic, or Newtonian paradigm. This paradigm is based on the doctrines of reductionism and universality, and emphasizes objectivity and the analytic mode of thought. Reductionism is based on the assumption that the most effective way to generalize is *from the parts to the whole*. In particular, it is believed that an understanding of the 'fundamental building blocks of nature' will allow us to grasp the 'natural laws' that govern both the universe and human behaviour. Analysis is the mode of thinking that complements reductionism. It deliberately attempts to break problems down into smaller sub-problems that are then treated as if they were *independent* problems. The assumption is that when all the sub-problems have been solved, then the whole problem will be solved. The part, or sub-problem, is the focus, and the relationship of the part to other parts and to the whole is typically ignored (Ackoff 1974).

Another doctrine is that universal (i.e. context-free) laws of nature exist 'out there', and that we can discover them only when our research environments are isolated, through rigorous controls, from the extraneous factors (contexts, or 'noise') that confound our results. A final doctrine of the Analytic Paradigm is that of objectivity: the belief that the practice of science is – or can be – value-free or value-neutral; that is, that scientists should be concerned not with what is right or wrong, or good or evil, but only with what is true or false. When a scientist loses this disinterest (e.g. is affected by the moral and ethical dimensions of his or her work), then s/he is no longer considered to be a scientist (cf. Bierstadt 1957: 10; Monod 1971: 13).

The practice of science historically has been based on unidisciplinary principles. This should not be surprising, because even the way universities are organized reflects the analytic mode of thinking: they are broken down into academic disciplines. This approach has demonstrated great utility in its application to well-structured problems that fall within the boundaries of a given discipline (Churchman 1971). However, those who have been trained in the disciplinary approach are largely unprepared by that training to address problems (or sets of problems) that transcend disciplinary boundaries. Nevertheless, the analytic approach is commonly applied to the conceptualization of transdisciplinary problems, even though it is, by definition, unsuited to such problems (cf. Mitroff and Sagasti 1973).

The reason for the continuing mismatch between problem type and conceptual approach is simple: there has traditionally been no alternative, and, as in the old saying, 'When the only tool you have is a hammer, you treat everything as if it were a nail.' This mismatch between the problem and the approach to its conceptualization has resulted in partial conceptualizations of world problems, and, hence, piecemeal and *ad hoc* approaches to research, and, therefore, ineffective solution strategies (Whyte 1982; Mitroff and Featheringham 1974).

The point is that the mode of thought that gave rise to the present world crises is unlikely to be effective in generating their solutions. Albert Einstein recognized the pressing need for an expanded approach to the conceptualization of problems as early as 1954, when he said, 'The problems that we have made as a result of the level of thinking we have done thus far cannot be solved at the same level as the level that created them' (Einstein 1960: 105). There is substantial agreement that problems that transcend disciplinary boundaries necessarily require a *trans*disciplinary, *integrative* approach to their conceptualization.

2.2 The integrative paradigm

Recently, however, a new paradigm has emerged that is more appropriate to conceptualize problems that are global in scale. This paradigm is based on the doctrines of expansionism, relativity and value-critical subjectivity, and emphasizes the integrative mode of thinking. Expansionism flows from the assumption that the most effective way to generalize is *from the whole to the parts*. This doctrine is expressed in the belief that all processes are inter-related, that is, that each part affects the whole system, and is in turn affected by the other parts. Because each part exerts a *non*-independent effect on the whole, the whole cannot be understood by analysing the parts independently. Thus, *the whole is always greater than the sum of its parts, because any system always exhibits some properties or behaviour that none of its parts can*. Explained in problem-solving terms, expansionism maintains that all problems can be viewed as subsets of larger wholes.

The integrative, or synthetic, mode of thinking complements expansionism. A fundamental principle of the integrative paradigm is that, when viewed structurally, a given problem (or set of problems) may indeed be divisible into subproblems; but when viewed functionally, it necessarily becomes an indivisible whole. The integrative approach is thus involved in synthesis in addition to analysis; instead of conceptualizing that which is to be explained (e.g. ill-health, poverty, high fertility, etc.) as an independent problem, it is conceptualized as an interdependent part of a larger problem set, and is explained in terms of its functional role in that larger system. This mode of thinking attempts systematically to connect problems with each other in order better to understand them by their interrelationships, which then are used as the basis of their resolution.

The doctrine of relativity holds that there are no universal (i.e. non-contextual) laws or truths, but only context-dependent processes. This doctrine is grounded in metaphors generated by the work of Einstein and Heisenberg (cf. Bohm 1980; Prigogine and Stengers 1984; Capra 1982), who found that Newton's allegedly 'universal' laws are in fact contextually determined. In the application of science to social problems, this means that contexts must be given priority consideration, not ignored or eliminated. A further implication of this doctrine is that scientists practise their craft in a context; and because different viewers' contexts necessarily differ (e.g. owing to disciplinary training), it is unlikely that any one viewer will be able accurately to perceive the whole (like the old Sufi fable of the blind men and the elephant).

Finally, the doctrine of value-critical subjectivity holds that humans are purposive, and hence all human practice, including the practice of science, is fundamentally normative and value-laden. All research serves human ends and purposes, and is therefore inescapably instrumental; this means that neither the problem under study nor the methodology applied is independent of either the researcher or the researcher's values (cf. Ravetz 1971; Rein 1976). Hence the integrative paradigm maintains that scientists cannot avoid being influenced – consciously or unconsciously – by what *should* or *ought* to be in their search for what *is* (i.e. what is true or false).

In order to ensure that the values inevitably linked to the scientific enterprise are made explicit, a 'value-critical' approach to the formulation of research problems and the implementation of research programmes becomes necessary. This aims to expose for critical review the values, ideals and ideologies that underlie competing perspectives regarding the problem under study (Rein 1976). Hence, the integrative paradigm includes the ethical dimension and its implications, which the analytic paradigm ignores.

3 CONCEPTUALIZING THE SAME PROBLEM FROM DIFFERENT PARADIGMATIC PERSPECTIVES

It was said earlier that the paradigm we apply to the problem is inevitably reflected in the way we *define* the problem. The choice among possible conceptualizations, or definitions, determines how the problem will be treated. In a very real sense, *the process of defining a problem is identical to the process of finding its solution* (Rittel and Webber 1973). And the application of different paradigms to the same problem results in very different problem definitions and, therefore, very different approaches to their solution. This principle can best be demonstrated by providing a concrete example; in this case we shall examine how differently 'health' is defined depending on whether one uses the analytic or integrative paradigm.

For the better part of this century health has been defined operationally (to guide activity in the field) as 'the absence of disease' (cf. Ratcliffe 1985) and this definition continues to provide the basis for the medical model structure of most health sectors today. The analytic, reductionist nature of this conceptualization is the basis of both health research and solutions to the problems of ill-health. First, illness and disease are perceived to be the result of a process that occurs at the level of the organism and the *individual*. Illness happens to the individual person because of specific conditions present in that particular individual, such as the presence of an infectious disease organism (e.g. bacterium), uncontrolled cell growth (tumour), genetic make-up, or personal life-style that make him or her *different* (abnormal, deviant) from everyone else.

From the perspective of this conceptualization, health problems are seen as consequences of individual human organisms interacting with disease organisms. Hence solutions focus on individual life-style change through such 'health promotion' techniques as jogging, and through *remedial* treatment, e.g. medication, chemotherapy, surgery, etc. Research activities informed by this definition focus on 'basic' research into disease mechanisms (e.g. cancer, AIDS) and on such technical and technological ways to treat and prevent disease and illness as genetic engineering, development of vaccines, finding means to eradicate disease organisms, and developing new medications and surgical procedures for treatment. In sum, health is seen as an individual problem, and solutions are seen primarily as technical and technological – and apolitical – in nature. From this point of view, the terms 'health' and 'health *care*' are seen as synonyms, and are used interchangeably.

But health can also be defined as 'a state of complete physical, mental and social well-being, and not merely the absence of disease or infirmity' (World Health Organization 1958). Two fundamental assumptions underlie this conceptualization (cf. Ratcliffe 1985: 96). The first is that the physical and mental health of human populations is inseparable from social health; and the second is that individual health cannot be understood (or treated) if it is viewed as separate – or separable – from the sociopolitical and economic system in which s/he is embedded and interacts from birth to death. From this perspective, all human beings are viewed as being *similar* to rather than different from one another. Solution strategies therefore tend to emphasize the *external* causes of health and disease patterns among *populations*, not individuals. That is, illnesses are seen as occurring with a certain probability within a given population grouping (an ethnic or occupational group, a social class, etc.) of persons who are in most respects like all others who are affected by the same environmental stressor (a mosquito-breeding swamp, a carcinogenic industrial pollutant, racial or gender discrimination, occupational stress, high unemployment rates, a persuasive advertising industry that promotes unhealthy products and/or life-styles, etc.) (Ryan 1981). This conceptualization recognizes that illness may indeed become manifest at the individual level, but that the causes of ill-health can occur at different system levels; hence prevention of ill-health requires intervention at all levels of the system, only one of which is treatment. It links observed disease patterns to social policy by maintaining that the unequal distribution of social goods and resources is the underlying cause of the differential disease patterns commonly observed between different social groups and classes.

Research informed by this definition focuses primarily on interrelationships between illnesses in populations and their social contexts, which are the result of particular patterns of social policy, and hence potentially preventable. From this perspective, it is more important to ask what kind of person has a disease than what kind of disease a person has. The emphasis is on social change as a preventive health intervention, rather than on individual change and remedial health interventions. In sum, health is viewed as a collective problem, and the required solutions are seen to be social and political in nature. In this perspective, the terms 'health' and 'health care' are not viewed as synonymous, and cannot be used interchangeably.

Clearly, those who define the same problem in different ways also structure their enquiry into that problem in very different ways, and will thus arrive at very different outcomes. As Kolata (1986) remarks, 'Some researchers are using molecular genetics to devise highly accurate tests of heart disease risk; others are trying to change the forces in society that put people at risk in the first place.'

4   HOW TO CHOOSE BETWEEN DEFINITIONS?

When even the experts disagree and support their positions with conflicting research evidence, how are we to choose between definitions that possess both apparent validity and internal consistency? The answer appears to be by reference to *meta-evaluations*, which are comprehensive reviews and assessments of the bulk of available research evidence relevant to the contending definitions of the same problem. The key to meta-evaluations is that the solution strategies (policies and programmes) dictated by an incorrect problem definition will not in fact solve the problem, and only a critical review of the research evidence *as a whole* will discover this fact.

Meta-evaluations of health-related research (cf. Cereseto and Waitzkin 1988; McKeown 1976; McKinlay and McKinlay 1977; Mosley 1983; Ratcliffe 1985; World Bank 1975) are unequivocal in their conclusions: the overwhelming majority of observed improvements in health (increased life expectancies, lowered infant mortality rates, etc.) have been brought about by public health measures, sanitation, and social policies that have resulted in a more equitable distribution of those basic resources essential to basic human security and survival. Despite the fact that some 95 per cent of health expenditures around the world are spent on curative services, it has been estimated that less than 10 per cent of observed declines in mortality in this century has resulted from such interventions (Ratcliffe 1985: 109; World Bank 1975: 256).

5 APPLICATION STRATEGIES OF THE INTEGRATIVE, TRANSDISCIPLINARY RESEARCH PARADIGM

The integrative paradigm provides instead the necessary holistic, multi-level conceptualizations of global problems that allow research and solution strategies to be mounted in an integrated fashion on several different system levels at the same time. This fact implies that there is the need for a fundamental shift in perspective not only at the abstract level but also at the concrete level – a paradigm shift means we not only must view traditional problems from a new perspective, we also need to rethink the traditional mode of conducting research.

Some of the most obvious organizing principles of integrative, transdisciplinary research methodology are detailed briefly in the following sections.

5.1 Placing health in a context

To conduct research in the integrative mode – viewing that which is to be explained as a part of a larger system, and explaining it in terms of its role in that larger system – thus requires, first of all, placing the problem in a context. An integrative research attitude would view society as both constructed of sub-systems (individuals, families, social groupings, classes, etc.) and itself as comprising a national system within a larger, geopolitical, regional system within a larger, global system. Therefore, it is recognized that research and conceptualization *must* take place at *all* system levels. As a first dimension, health-related research, for example, can be undertaken at the symptom level; shifting the focus to other levels of the system will not only result in increased complexity but will also lead to a larger number of options for long-term action. For example, application of the integrative approach to health-related research reveals at least four different system levels.

Level 1: Symptoms

The directly observable manifestations of ill-health (disease, illness and mortality) are found at the level of the individual in a society (e.g. dysentery, malnutrition, cardiovascular disease, TB, cancer, AIDS, violent deaths, accidents, etc.). Research and recommendations at this level tend to focus on disease mechanisms, immunological mechanisms, and physical and biological agents. It is difficult to base any action other than treatment of the individual and immunization of population groups on the basis of research on these symptoms alone, although the symptoms do indicate that the causal factors exist at a higher system level.

Level 2: Pre-symptomatic state

At this level of the system are those individuals and social groups who are 'at risk'. For example from poverty and chronic undernutrition; belonging to a minority group and suffering discrimination; unhealthy life-styles; drug abuse (including smoking, alcohol, prescription drugs); living in areas where air, land, food and/or water are polluted; homosexuality; etc., all of which can result in compromised resistance among these groups, and hence higher incidence and prevalence of illness, disease and death. Research and recommendations at this level tend to focus on secondary prevention (e.g. feeding programmes and food ration shops, needle exchange programmes, health promotion, etc.), although the problems at this level appear also to be symptoms of causal factors at a higher system level.

Level 3: Underlying causal correlates

The underlying causes of ill-health, disease and mortality in a society are many and complex. But many of the factors are a result of the maldistribution – through social and economic policies – of social 'goods' (i.e. resources that are essential to basic security and survival of humans, like employment, housing, basic welfare and health services, education, and unpolluted air, food and water) and of social 'bads' (i.e. pollutants, destruction of forest lands, repression of minority groups, organized conflicts with other nations over resource exploitation, and the pervasive and persuasive advertising industry and their benefactors, the powerful manufacturing industries that produce products that promote ill-health and death, e.g. alcohol, tobacco, chemicals, weapons, etc.). The production, distribution, and therefore consumption of all these goods and services are determined by the socioeconomic structure of the society, including its political and ideological structures. Hence research and resulting strategies tend to focus on these areas, and on revolution and/or reform of these structures and of the policies to which they have given rise. However, problems at this level can still be viewed as symptoms of problems occurring at a higher level in the system.

Level 4: Basic causal correlates

The conflicts and the interrelationships within the world political economy and between national economies and political and ideological structures are the ultimate determinants of international ill-health and mortality. The basic causal correlates explain how the potential resources of the international community of nations are mobilized for the production of goods and services, and how these are distributed among competing groups and classes, wars and revolutions, exploitive governments and multinational corporations, superpower competition for political and economic dominance, and geopolitical and economic injustices. Research and resulting strategies tend to focus on these areas (e.g. how pressures exerted by more powerful nation-states on weaker nations divert resources away from the poor and weak to the rich and powerful – the net outflow of $33 billion in 1988 from the 'underdeveloped' South to the 'overdeveloped' North provides but one example), on the interrelationship between the global geopolitical economy and the health of the natural ecosystem on which all of humankind is dependent for survival, and on the need for revolution and/or reform of malignant social and economic forces at the international level.

5.2 Linking levels in the system

In order to understand health fully, one must understand the links between the different levels of the system within which ill-health, disease and mortality are defined as problematic. Most research on such global problems has focused on studies at the levels of symptoms and immediate causal correlates. For example, in studying malnutrition we know much more today about the metabolism of nutrients and the processing dimensions of food than we do about the concrete role of ideology and politics in creating, alleviating, or preventing malnutrition in society. Yet we know that interventions at the symptom or presymptomatic levels are unlikely to solve the problem; in most instances, such problems respond only to changes in underlying and basic causal correlates (cf. Mosley 1983; World Bank 1975). That is why studies at the levels of basic and underlying causal correlates are so important today.

Research with an integrative approach is necessarily transdisciplinary, while problems examined from a disciplinary perspective only lead to very narrow solutions. For example, physicians tend to recommend only medical interventions, even where the evidence shows that health levels are determined primarily by public policies that differentially distribute basic social 'goods' and 'bads' among competing social groups and classes (Brenner 1976a, 1976b, 1979; Dubos 1971; Illich 1975; Mahler 1976; McKeown 1976; McKinlay and McKinlay 1977; Mosley 1983; Ratcliffe 1985; World Bank 1975). While such narrowly conceived reductionistic approaches and linear-causal interventions may be appropriate in specific instances, they are typically incomplete and often counterproductive. Poverty, illiteracy, unemployment, high fertility and mortality, inadequate housing, and hunger at the societal level are *system* characteristics (cf. Ackoff 1974; Churchman 1979; Einstein 1960), and a transdisciplinary, task-force approach is essential to the full understanding of such complex problems and their interrelationships.

5.3 The transdisciplinary, task-force approach to research

First we must clarify the difference between *trans*disciplinary research and *multi*disciplinary research, with which it is commonly confused. Multidisciplinary research typically exhibits three main characteristics: one is common agreement on the importance of a particular problem to be solved; another is that the research team represents several different disciplines; and the third is the binding assumption that the sum of the disciplinary approaches of the team members – each to his or her own particular aspect of the problem – will provide a holistic conceptualization of the problem. Such teams also tend to be pluralistic in values and interests, although the impact of these differences on the research process is not explicitly addressed. While this is an improvement upon the unidisciplinary approach that typifies the analytic paradigm, it does not constitute a transdisciplinary approach to research.

A transdisciplinary team is also problem-oriented, but its members either share views regarding the basic nature (i.e. definition) of the problem under examination or they try to discover why their views differ. If agreement regarding the basic nature of the problem cannot be reached, then the research effort investigates possible solutions based on more than one definition of the problem (cf. Mason 1969). But the binding factors are shared value-commitments to integration and an attempt on the part of each member to transcend his/her own disciplinary perspective to arrive at a *shared* perspective that will provide a less partial conceptualization of the problem. This allows each to place their own particular aspect of the whole problem in its proper context: as one interdependent element of the whole.

The transdisciplinary, task-force approach is necessary to delve into the various geographical levels of the international system at which the problem may appear, for example, malnutrition due to inadequate intake of food at the individual level can be taken as a symptom of the problem caused by maldistribution of food (underlying cause). But the basic causal correlates of maldistribution of food in society may exist at many different levels of the world system. Table 12.1 gives an example of a logical scheme in which such a global health problem can be shown to arise at a number of different levels.

This integrative and transdisciplinary approach to the problem of hunger demonstrates that underlying causal correlates may be identified at any geographical level, while basic causal correlates are typically limited to the national and international levels. It therefore becomes clear that the promotion of health and the alleviation and prevention of ill-health require concerted action at each and every system level to which causal correlates can be traced. Yet the traditional, analytic approach to research is characterized by individuals or teams from a single discipline mounting uncoordinated, *ad hoc* research efforts without consideration of other dimensions of the problem and the needs of other researchers. Indeed, the way analytic research has

Table 12.1 Correlates of global health

	Historical causal correlates	Current causal correlates				
		Political	Economic	Ideological and cultural correlates	Informational	Ecological/technological
International level (Between nations)	Historical evolution of technological, geographical, ideological, cultural, economic, and political power relationships in the global context, etc.	Distribution of political, economic, and military power relationships as it affects war, intimidation, access to international decision-making, freedom and democracy, etc.	Economic power relationships; centre-periphery relations; imperialism/ colonialism/neocolonialism; international division of labour as it affects distribution of social and natural resources, etc.	Power relationships (spheres of influence) of differing ideologies, cultural values, etc., as they affect patterns of distribution and consumption of world resources, etc.	How differential access to information and knowledge affects power relationships and, thus, patterns of production, distribution and consumption of international resources, etc.	Relationship between technology and ecology (e.g. exploitation/conservation/ materialism/humanism) and effects of ethics on production, distribution, consumption, etc.
National level (Between areas, regions, districts, etc.)	Evolution of governmental type and technological, geographical, religious, economic, and political power relationships over time, etc.	Power relationships as they affect human rights, access to political decision-making process (democratic/totalitarian), distribution of social resources, access to legal resources, etc.	Centre-periphery power relationships as they affect access to means of production, labour relations, and distribution of production, consumption, health, fertility, etc.	Espoused and actual ideological ethic(s) (political, religious, economic) as they affect power relations and resource distribution between social groups/classes, etc.	How power relationships affect the distribution of information and knowledge within society and thus patterns of resource distribution, etc.	Power relationships in terms of access to/control of others' resources via technological (industrial/ military) superiority and ecological ethic.
Local level (Between villages)	Evolution of sociopolitical, economic, technological and religious power relationships as they affect production, distribution and consumption over time, etc.	Power relationships in terms of access to political decisions affecting area; fiscal/legal control; democratic/ authoritarian, etc.	Power relationships as they affect ownership or access to means of production, distribution, consumption; competitive/co-operative, etc.	How ideological and cultural ethic(s) affect power relationships and patterns of social resource distribution and consumption, etc.	How distribution of information affects power relationships and, thus, patterns of production, distribution, and consumption, etc.	Differential access to/control of own/others' resources, land, fertility, distribution of technical knowledge, etc.
Village level (Between households)	Evolution of the distribution of economic, religious and technological power, and the resultant distribution of social resources, etc.	Power relationships in terms of access to political decisions affecting village; fiscal/legal control; democratic/authoritarian, etc.	Power relationships as they affect ownership or access to means of production, distribution, consumption; competitive/co-operative, etc.	How power relationships and hence access to social resources are affected by political, economic and religious ideologies, etc.	How distribution of information affects power relationships and hence patterns of production, distribution and consumption, etc.	Evolution of the distribution of technological knowledge and its relationship to ecological ethic, production, distribution and consumption, etc.
Household level (Between household members)	Evolution of structure, role functions of family members, division of food and labour, access to means of production, educational distribution, etc.	Power relationships within family by age, gender, etc. as they affect access to family decision-making process, status and role-functions, education, nutrition, etc.	Ownership or access to means of production; division of labour by gender, age, etc.; economic distribution; participation in family economic decisions, etc.	Ideological and cultural values as they affect the distribution of and access to social resources (education, food, finances, etc.) within the family.	How distribution of information and knowledge affects patterns of access to and distribution and consumption of resources, etc.	Soil, climate, natural resource base, access to means of production, division of labour, distribution of technical knowledge, etc.

Source: Adapted from United Nations University (1981), 'Hunger and Society; Sub-Programme 1 of the World Hunger Programme', Report of a UNU World Hunger Programme Ad Hoc Working Group, Feb. 16–19, Tokyo: The University, p. 5.

historically been conducted is analagous to a large number of highly trained musicians playing different instruments in a totally unco-ordinated way, in different keys, and in different time. The transdisciplinary, task-force approach, on the other hand, is analogous to a symphony, where the musicians agree to merge their different skills and different instruments in a concerted effort to achieve that which none can achieve separately.

Those familiar with health-related research will see that most studies in the field are analytical and partial, and that they tend to focus primarily on symptom correlates rather than immediate and/or basic causal correlates. For this reason, analytic, disciplinary research typically focuses on the *consequences* of ill-health and recommends *remedial* action that has the potential only for alleviating disease and ill-health on an *ad hoc* basis at the level of the individual. Integrative, transdisciplinary research, on the other hand, tends to focus on *causal correlates*, and recommends policy interventions at higher system levels aimed at *promoting* health and *preventing* ill-health.

6 CONCLUSION

Although the terms 'method' and 'methodology' are commonly used interchangeably, the terms are not equivalent. 'Method' denotes a scientific procedure; it is the specification of the steps required (the means) in order to achieve a desired end (e.g. to collect or analyse data). Scientists are usually familiar with at least some of the assumptions that underlie specific *methods* (e.g. normal distribution of data), and those that distinguish particular *classes* of methods (e.g. qualitative and quantitative). 'Methodology', on the other hand, refers to the entire system of philosophical assumptions (articles of faith) about humans and the world that one holds so deeply to be true that they are never even questioned; this includes assumptions about epistemology, about thought, about logic and about science, as well as about theory and method. Methodology is not a term that is well understood among the scientific community, and this lack of understanding is a major constraint on our ability to conceptualize world problems in such a way as to allow their effective resolution.

The point is that our ability to manage our complex affairs and solve human problems depends more on our understanding of and attitudes toward the world – i.e. our paradigm – than on science or technology. To put it another way, successful problem-solving requires finding the right solution to the right problem. Yet we fail more often because we attempt to solve the wrong problem than because we apply the wrong solution to the right problem – apparently because the mode of thinking that created these problems is incapable of solving them. And if we are to resolve the problems that face us today, we need to change our mode of thinking.

The series of interrelated global crises we face today threatens us with early deaths as individuals, and with extinction as a species. Attempts to understand and resolve these problems through application of the analytic, unidisciplinary approach to scientific problem-solving have only served to demonstrate that this approach is unsuited to deal with crises of this magnitude. However, it is said that the Chinese character for 'crisis' is a combination of two other characters: 'danger' and 'opportunity'. In this instance, the *danger* is that the uni-disciplinary enquiry system in which we have been trained and with which we are comfortable will continue to determine how the problem will be approached. Should this occur, then we shall fail to resolve these crises because we continue to apply a scientific paradigm and research methods that are mismatched to the problems that command our attention. The *opportunity* is that the potential exists to resolve these crises through the application of a scientific paradigm and research methods that are matched to the problem class under consideration. It is up to those scientists who care about individual health and species survival enough to transcend their disciplines and work together to understand and resolve these international crises.

REFERENCES

Ackoff, R. L. (1974) *Redesigning the Future*, New York: Wiley-Interscience.

Bierstadt, R. (1957) *The Social Order*, New York: McGraw-Hill.

Bohm, D. (1980) *Wholeness and the Implicate Order*, London: Routledge & Kegan Paul.

Borlaug, N. E. (1986) 'Accelerating agricultural research and production in the Third World: a scientist's viewpoint', *Agriculture and Human Values* 3(3): 5–14.

Brenner, M. H. (1976a) 'Mortality, social stress, and the modern economy: experience of the United States, Britain, and Sweden, 1900–1970', paper presented at the Annual Meeting of the American Association for the Advancement of Science, Boston, 12 February.

—— (1976b) *Estimating the Social Costs of National Economic Policy: Implications for Mental and Physical Health and Criminal Aggression*, US Congress, Joint Economic Committee, Washington, DC: USGPO.

— — (1979) 'Mortality and the national economy: a review and the experience of England and Wales', *Lancet*, 15 September.

Brown, L. (1984, 1985, 1986, 1987, 1988) *The State of the World* (annual publication), Washington, DC: World Resources Institute.

Capra, F. (1982) *The Turning Point*, New York: Bantam.

Cereseto, M. and Waitzkin, H. (1988) 'Economic development, political-economic system, and the physical quality of life', *J. Public Health Policy*, Spring, pp. 104–20.

Churchman, C. W. (1971) *The Design of Inquiry Systems: Basic Concepts of Systems and Organization*, New York: Basic Books.

—— (1979) *The Systems Approach and Its Enemies*, New York: Basic Books.

Dubos, R. (1971) *Mirage of Health; Utopias, Progress and Biological Change*, New York: Harper & Row.

Einstein, A., in O. Nathan and H. Norden (eds) (1960) *Einstein on Peace*, New York: Schocken Books.

Eisenberg, L. (1977) 'The perils of prevention: a cautionary note', *New England J. Medicine* 197: 1213.

Illich, I. (1975) *Medical Nemesis; the Expropriation of Health*, London: Marion Boyars.

Knowles, J. (1977) 'Responsibility for Health', *Science* 198: 1103.

Kolata, G. (1986) 'Reducing risk: a change of heart?', *Science* 231: 669.

Kuhn, T. (1970) *The Structure of Scientific Revolutions*, Chicago: University of Chicago Press (2nd edn).

McKeown, T. (1976) *The Role of Medicine: Dream, Mirage, or Nemesis?*, London: Nuffield Provincial Hospital Trust.

McKinlay, J. B., and McKinlay, S. M. (1977) 'The questionable contribution of medical measures to the decline of mortality in the United States in the twentieth century', *Milbank Memorial Fund Quarterly*, Summer, pp. 405–28.

Mahler, H. (1976) 'A social revolution in public health', *WHO Chronicle* 30: 475–80.

Mason, R. O. (1969) 'A dialectical approach to strategic planning', *Management Science* 13, pp. B403–B414.

Mitroff, I. and Featheringham, T. R. (1974) 'On systemic problem solving and the error of the third kind', *Behavioral Science* 19: 383–93.

Mitroff, I. and Sagasti, F. (1973) 'Epistemology as general systems theory: an approach to the design of complex decision-making experiments', *Philosophy of the Social Sciences* 3: 117–34.

Monod, J. (1971) *Chance and Necessity: An Essay on the Natural Philosophy of Modern Biology*, New York: Knopf.

Mosley, W. H. (1983) 'Will primary health care reduce infant and child mortality? A critique of some current strategies, with specific reference to Africa and Asia', International Union for the Scientific Study of Population (IUSSP) Monograph, Inst. National d'Etudes Démographiques, Paris, March.

Prigogine, I. and Stengers, I. (1984) *Order Out of Chaos: Man's New Dialogue with Nature*, San Francisco: W. H. Freeman.

Ratcliffe, J. W. (1983) 'Notions of validity in qualitative research methodology', *Knowledge: Creation, Diffusion, Utilization* 5(2): 147–67.

—— (1985) 'The influence of funding agencies on international health policy, research and programs', *Mobius* 5(3): 93–115, July.

Ravetz, J. R. (1971) *Scientific Knowledge and Its Social Problems*, New York: Oxford University Press.

Rein, M. (1976) *Social Science and Public Policy*, New York: Penguin, p. 140.

Rittel, H. W. J. and Webber, M. (1973) 'Dilemmas in a general theory of planning', *Policy Sciences* 4: 155–69.

Robinson, J. (1964) *Economic Philosophy*, New York: Penguin.

Ryan, W. (1981) *Equality*, New York: Vintage Books.

Sivard, R. L. (1986, 1987, 1988) *World Military and Social Expenditures* (annual publication), Washington, DC: World Priorities.

United Nations University (1981) 'Hunger and society; sub-programme 1 of the World Hunger Programme', *Report of a UNU World Hunger Programme Ad Hoc Working Group*, 16–19 Feb., Tokyo: United Nations University.

Whyte, W. F. (1982) 'Social inventions for solving human problems', *American Sociological Review* 47: 1–13.

World Bank (1975) *Assault on World Poverty: Problems of Rural Development, Education, and Health*, Baltimore, Md: Johns Hopkins University Press.

—— (1984, 1985, 1986, 1987, 1988) *World Development Report* (annual publication), Washington, DC: World Bank.

World Health Organization (1958) Constitution of the World Health Organization, annex F, in *The First Ten Years of the World Health Organization*, Geneva.

13

CAN HEALTH BE A QUANTITATIVE CRITERION? A MULTI-FACET APPROACH TO HEALTH ASSESSMENT

David Canter and Lorraine Nanke

1 WELL-BEING AND ILLNESS

There appear to be two separate but overlapping realms of discourse in the consideration of health. One area deals with the experience of well-being and competence. Within the psychological literature it often includes such issues as self-esteem, satisfaction with living and working conditions, sense of autonomy, and other aspects of day-to-day experience. Discussions about the causes and consequences of different states of well-being are often cast in a social, psychological or sociological framework. Relationships with others, effective support from other people and understanding of current circumstances, are all aspects of studies of well-being. Threats to well-being are also part of the consideration of interpersonal processes. Stress can be viewed as a strong challenge to well-being. Psychologically, the reductions in well-being are seen as beginning with stress and ending in unhappiness, depression and despair.

Close to those factors concerned with happiness and well-being lie those whose primary focus is the health of the individual. As has often been pointed out, most researchers who study health do so by studying the absence of illness and disease. Their primary concern is with the invasion, decay and destruction of the functioning components of the physical organism. Individual health is therefore often described in terms of mechanical damage, chemical interference and biological change.

Although the realm of well-being and the realm of individual (biological) health status appear to be facing in opposite directions their attention is actually pointing to the common object of a person's experience of the effectiveness of his or her own personal functioning. Any progress in bringing together the realms of well-being and illness therefore requires forms of assessment of the personal experience of sub-optimum functioning. Any such assessment will need to cover the social, personal and physical manifestations of an individual's experiences.

183

2   FACETS OF TIME AND PLACE

A person's experiences are defined by the socio-cultural matrix of which he or she is a part. The same physiological experience may be interpreted in quite different ways in different contexts. For example studies of the placebo effect show that the interpretations which people make of what is happening to them are influenced by their expectations, these expectations in turn being a product of previous situations that the person either knows about or has had direct contact with. A milestone in the experience of illness or well-being is the point where distortions in what is assumed to be normality are found to be so challenging that the individual seeks help.

At the moment of treatment the various issues associated with health are brought into sharp focus. The medical practitioner has to try and conceptualize the expressed experiences of the patient within a particular therapeutic framework. However, even here the practitioner will attempt to find out by whatever means available the likely antecedents to the current state. In general, where health and physical fitness are concerned, although various bio-physical assessments will be relevant, the dominant focus of concern stems from the patient's own account of his or her experiences.

We are proposing, therefore, that any assessment of health and any attempt to measure its various aspects may fruitfully begin with the careful exploration of the account which the patient gives at the time of seeking treatment, although it should not be forgotten that descriptions of health arise from someone who feels the need for treatment. As will become clear, such a focus inevitably has a multi-faceted framework. The idea that health has many facets is technically quite a precise concept. It is based on the recognition that any account of experience is susceptible to different modes of assessment. Together these make up the overall subjective experience but for measurement and analysis the different constituents need to be established. Once established these constituents do, in effect, provide a theory of health experience.

This essentially multivariate approach to the assessment of health is rather different from the usual procedures that focus either on some physical symptoms, such as body temperature or white blood corpuscle count or on some focused aspects of mental state, such as those examined in questionnaires such as the General Health Questionnaire (Goldberg 1978). Perhaps the most obvious facet of health is that it can be thought of either in terms of the negative qualities of disability and illness or in terms of the positive qualities of physical fitness and sense of well-being. A second facet of accounts of health experience, also inherent in our earlier discussion, is that they all do have a strong temporal component. The experience of the present is shaped by what is known of the past and what is expected for the future. Therefore we should consider the patient's view of the longer-term consequences of attempts to alleviate symptoms as well as the more recent experience (say in the past week) that has preceded the current consultation. This brackets the patient's current conceptualizations of their state.

A third component may be regarded as the anticipation of the likely effects and benefits of treatment or of the natural unfolding of the illness. There is a complication in considering anticipations and expectations of an illness as a separate sub-category in the temporal sequence. A person's prediction of the future will be greatly influenced by the person's past perception of the rates and types of change that they have already been experiencing in their symptoms. Therefore the disabilities currently experienced may be a result of the patient's perception of how long they have been suffering from that particular problem.

Besides the self-assessment of health, and the temporal dimensions, there is a further facet that is needed to provide a more detailed understanding of the patient's experience. This draws attention to the reference frames or foci of the experience. Putting them simply, these foci may be thought of as physical, emotional or social. In the physical foci, issues of pain and feeling ill, as well as the positive issues of physical fitness, will be important. The emotional aspects of the experience will overlap with these but may be characterized by feelings of energy or happiness at the positive end, or more general descriptions of tiredness or feeling ill on the negative side. The social foci relate to the person within a social context. Inability to cope with other people may be a reflection of general distress or other aspects of the experience of illness. In practice, this element is often seen as connecting to problems of coping with work or maintaining work activities.

3 A GENERAL MAPPING SENTENCE FOR HEALTH

These three aspects, or facets, of the experience of health are seen as interrelated constituents of the same system, rather than separate dimensions or isolated clusters. They are different ways of looking at the same process. Their particular meanings and significance will be derived from their relationships to each other. It is interesting that whenever we ask a question of a patient we implicitly select from one of the three facets. We address either illness or well-being; we should be aware whether it is the present, the past or the future that is being considered, or whether it is the physical, emotional or social aspects of the experiences. Of course, various levels of ambiguity are possible. The question 'How are you today?' is ambiguous in terms of illness or well-being and the particular reference to which attention is being drawn. But it is quite clear that it is focusing on the present in contrast, say, to the question 'How long have you had this problem?'

The facets can be listed in relation to each other with connectives, in what has been termed a 'mapping sentence' (Canter 1985). This is a simple summarizing device for encapsulating all the facets in relation to each other.

A general mapping sentence of health might be as follows:
The extent to which a person describes him/herself as being:

[physically] [ill]   [in the future   ]
[emotionally] [well]   [at present   ]   extremely ill
[socially   ]   [in the recent past]   →
                        [in the longer past]   well

This summary mapping sentence is a whole set of interrelated hypotheses about a patient's experience. The hypotheses can be grouped into three levels. At the first level it can be suggested that the elements of each facet are distinct aspects of the individual's experience. For example, that a person's answers to questions about their long-term problems will be distinct from the answers that they give to shorter-term problems, or that the physical symptoms will be distinguishable from the emotional symptoms which in turn will be distinct from the social problems with which they may be associated. Perhaps the most fundamental of this group of hypotheses, is that positive and negative aspects of health are not the same phenomena merely expressed in different words, but are two different processes that interact. One is the process of debilitation and illness. The other is the process of enrichment and life enhancement.

The second set of hypotheses concerns the relationships between the elements. For example, the physical experience might be conceived as relating to the emotional aspects. These in turn could be linked to the social and work-related issues. Within this essentially reductionist perspective the emotional aspects might be central, implying an order between the three main elements from physical on to social. Similarly the temporal facet has a basic logic implicit within it. The current situation would be expected to connect most closely to the most recent past and this in turn to the longer past. However, these are hypotheses that imply a particular psychological ordering of experience. Other possible orders between the elements are feasible. For example, the present may be seen as a review of long-term processes together with predictions of the future. All of these possibilities will be modified by the particular forms of illness or well-being that are being explored. So there is a very real empirical question as to what are the general trends across a particular set of data.

All these facets are part of the same system of health experience. Therefore, one could suggest that there are aspects of the patient's experience which are central to the whole process, which capture its essence. Take the analogy of a motor car. It has components with very different roles, such as the carburettor, engine and luggage compartment. Yet taken together these components could indicate some quality of the car such as its luxuriousness or power. In this analogy the power would be an essential quality of the overall system that captures aspects that are reflected in its more differentiated components. There can be additional hypotheses as to what the core issues may be, especially when we are exploring the structure of illness and well-being during the process of treatment. Is it the experience of physical fitness? Or the general feeling of illness? Or is it the particular role of the treatment process itself and the search for health?

4 AN EMPIRICAL TEST OF THE MAPPING SENTENCE

In practice, in order to test the validity of this mapping sentence each element of each facet can be combined together to define a template, profile, or as it is sometimes called, a structuple. Each of these structuples can be turned into a number of different questions. A questionnaire can therefore be developed which covers all the possible combinations of issues indicated by the mapping sentence.

The questionnaire can then be used to test and elaborate the mapping sentence by the consideration of the correlation that every question has with every other question for a sample of patients. In this process the patient's answers to two different questions are taken as an indication of whether or not those questions share some common psychological process. If they do, then when one question is answered in an affirmative direction it would also be expected that the second related question would also be answered in an affirmative direction and vice versa. A spatial representation of the correlations is useful for the examination of the overall pattern of correlations.

5 AN EXAMPLE OF THE USE OF THE FACET METHOD

A questionnaire was developed covering the issues described in the mapping sentence (see Figure 13.1). The fifty-two questions asked are given there. We will look at how this was used in one study of ninety-two patients attending a variety of homoeopathic and other complementary practitioners for treatment. The sample was somewhat *ad hoc* in that the patients were entered into the project at the discretion of the practitioners. They were participating in a larger-scale study of the efficacy of homoeopathy. The practitioners, only some of whom were medically qualified, covered a very wide range of types of practice, including iridology and reflexology as well as hypnotherapy, herbalism and various forms of manipulation. It was also clear that the patients suffered from a very wide range of medically diagnosed illness. The answers to the questions ranged from 'not at all' through to 'slightly', 'quite', 'very' and 'extremely'. These answers were scored from one to five and each of these numbers for the ninety-two patients provided the raw data for subsequent analysis.

6 SMALLEST SPACE ANALYSIS (SSA)

In order to examine the underlying structure of the correlations an algorithm developed by Guttman and Lingoes was used. This is known as Smallest

SYMPTOM INTENSITY QUESTIONNAIRE

Please tick the box which most nearly describes your experience of THE MAIN COMPLAINT for which you are seeking treatment OVER THE PAST WEEK.

(1) How well do you feel you understand what is wrong with you?

totally	mostly	quite a bit	very little	not at all

(2) How often have you suffered from this complaint?

never	occasionally	quite often	very often	all the time

(3) How severe has this complaint been?

not at all	slightly	quite	very	extremely severe

(4) How difficult has it been for you to control this complaint?

not at all	difficult	slightly	quite	very	impossible

(5) How much did this complaint interfere with your normal life?

not at all	slightly	quite	very much	completely

(6) How serious do you think this complaint is?

not at all	slightly	quite	very	extremely

(7) How would you rate your general health?

very good	good	average	poor	very poor

(8) How responsible do you feel for your current state of health?

totally	mainly	a bit	slightly	not at all

HOW LONG have you suffered from the following problems?

	doesn't apply	less than 1 month	1 to 6 months	6 months to 2 years	more than 2 years
(9) Your current main complaint					
(10) Feeling inappropriately tired					
(11) Feeling emotionally distressed					
(12) Difficulty coping with work					
(13) Difficulty getting on with people					
(14) Feeling generally ill					
(15) Your current physical symptoms					
(16) Your current pain					

(cont.)

How much have you experienced the following OVER THE PAST WEEK?

	not at all	slightly	quite a bit	very much	severely
(17) Feeling inappropriately tired					
(18) Feeling emotionally distressed					
(19) Difficulty coping with normal work					
(20) Difficulty getting on with people					
(21) Feeling generally ill					
(22) Suffering from physical symptoms					
(23) Being in pain					
(24) Feeling mentally alert					
(25) Feeling very happy					
(26) A strong sense of achievement					
(27) Enjoying being close to someone					
(28) Feeling physically fit					
(29) Feeling very energetic					
(30) Feeling deeply relaxed					

How important were the following problems IN YOUR DECISION TO SEEK TREATMENT NOW?

	not at all important	slightly important	quite important	very important	the most important
(31) Feeling inappropriately tired					
(32) Feeling emotionally distressed					
(33) Difficulty coping with normal work					
(34) Difficulty getting on with people					
(35) Feeling generally ill					
(36) Suffering from physical symptoms					
(37) Being in pain					

(38) How confident are you that this treatment can help?

completely	very	quite	slightly	not at all

(39) How many months have you been attending this practice?

1st visit	under 1 month	1–6	6–24	over 24 months

(40) Approximately how many minutes did you spend with your practitioner at the most recent consultation?

under 5 minutes	5–15	15–30	30–45	over 45 minutes

(cont.)

189

What influence do you think the following have had on your state of health over the past month?

	not applicable	worse	same	slightly better	much better	completely well
(41) Homoeopathic remedy						
(42) Homoeopathic consultation						
(43) Other alternative treatment						
(44) Orthodox medical treatment						
(45) Your own attitude						
(46) Your own behaviour						
(47) Changes in your life						
(48) Other [please specify]						

Over the past month, how would you rate:
(49) Changes in main complaint?
(50) Changes in general health?

How do you think your condition will change:
(51) Over the next 4 to 6 weeks?
(52) In the long term?

Figure 13.1 Symptom intensity questionnaire
Source: D. Canter and L. Nanke 1992

Space Analysis. As discussed in the Technical Appendix this is one of the family of multivariate procedures that attempts to establish the structure underlying a matrix of correlations. This algorithm starts from an initial correlation matrix. In the present case the correlation matrix was derived from the correlation every one of the fifty-two questions had with each other. In this case product–moment correlations are used.

The algorithm then represents the correlations between the variables as distances between points in a space. Finding the best, in this case two-dimensional, representation, the computer creates a space in which each point is one of the questions asked. The further apart the questions are from each other the lower the correlation between those questions. The visual representation, shown in Figure 13.2, therefore, summarizes the key relationship that every question has to every other question. This allows us to see the underlying structure and pattern within the original questionnaire. From examination of this structure it is possible to see which of the hypotheses enshrined in the mapping sentences are supported by actual data. A further advantage of the geometric representation is that it can facilitate the development of new hypotheses for subsequent testing.

7 INTERPRETATION OF THE SMALLEST SPACE ANALYSIS

The most fruitful approach to interpreting an SSA has been found to be the utilization of *regional hypotheses*. This proposes that items that incorporate any element of a mapping sentence should be together in the same region of the space. Because we are looking at a multi-faceted system there are a number of different ways in which the questions may be correlated with each other, and therefore, next to each other in space. A complex multidimensional structure is therefore feasible if all aspects of the mapping sentence are to be reflected in the data. For the present discussion, though, a two-dimensional representation of the relationships will be considered. This will enable us to see the dominant facets that are clearly reflected in the regional structure of the SSA space.

Because we are concerned with the relationship each variable has to every other there is no necessary significance to the horizontal or vertical axes in the space. Instead the regional areas of the space are the primary concern. If this regional structure implies horizontal or vertical interpretations then they may be of value, but no such axes are assumed *a priori* to be relevant.

The plot can be interpreted by considering sets of questions that are grouped together. Each point on the plot is represented by a number, which corresponds to the question listed by the same number in the questionnaire. The points representing the questions in Figure 13.2 reveal distinct sets of correlations that pull questions together into regions and distinguish them from other questions. The facet structure represented by questions is empirically supported to the extent that it corresponds to the regional grouping of items on the plot. Using this approach, the interpretation of the plot provides some support for the following regional hypotheses.

8 THE HORIZONTAL DIMENSION: GENERALIZED EXPERIENCE OF ILLNESS AND WELL-BEING

8.1 The domain of illness experience

Grouped together on the centre left of the plot are a set of questions which include physical descriptions of illness and their psychosocial sequelae. The

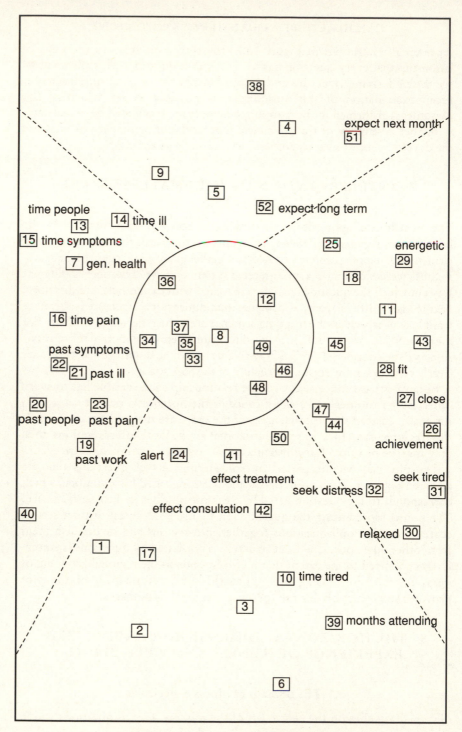

Figure 13.2 First symptom intensity SSA

group of items to the centre left of the plot includes questions 19, 20, 21, 22 and 23 (past week's experience of illness, symptoms, pain, difficulties at work and with other people), and questions 13, 14, 15, 16 and 19 (longer-term history of illness, symptoms, pain and difficulties with people) as well as question 7 (general health).

What is not apparent in this region is any distinction between the different referents of illness. Questions to do with pain or the difficulty of getting on with people do not form any distinct regions within this area of the space.

It would seem that the patient's experience of illness is a composite of all the physical and social referents, with the exception of emotional distress and tiredness. Referents do not appear to operate as distinct elements in this particular sample and analysis, so the region can best be described as history of physical and social debility, generically referred to as illness.

Considering the plot as a whole, items which fall within the left side include questions 9 and 2, dealing with the duration and frequency of the patient's main complaint, and question one concerning patients' understanding of their condition, suggesting that patients interpret these items in terms of their overall experience of illness.

8.2 The temporal structure of illness experience

The illness region is subdivided in terms of time. The lower section contains questions on the experience of different aspects of illness over the past week. The upper section contains items dealing with the length of time the patient has suffered from various aspects of illness, from less than one month to two years and more. This supports the hypothesis that both the past week and longer-term history provide distinct aspects of health experience. We can therefore see the temporal facet nested within the main perceptions of illness experience.

The data are consistent with the view that illness can be a reminder of mortality, so that temporal factors exert a more profound influence over perception. From a psychological perspective, subjective time varies according to the nature and type of experience. Unpleasant experiences like pain or illness constrict the sense of time, so that objectively short periods of time seem to last longer. It is particularly interesting that questions 17 and 24, concerning experience of tiredness and alertness during the past week, are located in the left side of the plot. The experience of tiredness or alertness may well influence patients' estimations of the history of their condition.

8.3 The domain of well-being

On the right-hand side of the plot are a group of questions dealing with emotional state, which roughly correspond to well-being and competence as described in the introduction. This cluster of items includes: (a) experience over the past week of being happy, energetic, distressed, fit, close to someone, of achievement, and relaxation; (b) the perceived influence on state of health over the past month of one's own attitude, life changes, orthodox and complementary medical treatment; (c) seeking help for tiredness and emotional distress; (d) longer-term history of emotional distress.

This region includes most of the positive well-being items, showing that these are distinct from questions dealing with illness. However, it is clear that these are not simply questions about the opposite to illness, because within this region there are other questions dealing with the recent and longer-term history of emotional distress, that might on the face of it have been thought to be part of the negative experience of illness. This suggests that there is a unique relationship between emotional debility (as opposed to physical or psychosocial debility) and general well-being.

The centrality of emotion in the structure of well-being is a particularly important finding for health psychology, which has tended to focus on cognitive, behavioural, social and physical aspects of health functioning, with comparatively little direct reference to emotion.

In other words, the grouping to the right of the plot deals with well-being, but focusing on emotional and more directly psychological aspects of that well-being. The overall focus can best be described as emotional, in contrast to the physical/ social focus of the left-hand cluster.

9 COMPARISON OF ILLNESS AND WELL-BEING REGIONS

9.1 Independent domains of experience

The sets of items dealing with illness and well-being items are quite distinct. The physical experience of illness and psychosocial sequelae are relatively independent of more emotionally focused experiences of well-being within the system of health experience. The distinction between these two sets of items is perhaps most readily highlighted by the two questions of a general nature that are in the centre of the two regions. In the left-hand region is the question 'How would you rate your general health?', and on the right hand is the question 'How long have you suffered from emotional distress?'

This challenges any simplistic 'healthy mind/healthy body' view of health, indicating that it is possible to experience oneself as healthy in either, neither, or both. Many people who are physically ill will exhibit a state of mental well-being, and many people who do not exhibit a state of mental well-being will be lacking in any signs of physical illness. There will inevitably be a relationship between physical and mental well-being in that physical illness is likely to be associated with lack of experienced well-being, but current findings suggest that this is a far from perfect relationship.

9.2 The usefulness of the generic terms 'illness' and 'well-being'

For patients in the current sample, the experience of illness does not appear to be differentiated into physical or social (i.e. difficulties at work or in relationships) components, though it does seem to be separate from emotional debility. Similarly, the positive well-being element of the facet on the right is associated with the physical, emotional and social elements on the left. In other words, the experience of well-being does not seem to be clearly differentiated into different foci or aspects of the self: there is a positive relationship between the different aspects of well-being. In both these cases, the plot provides some support for use of generic terms 'illness' and 'well-being'.

In terms of the original facet structure, illness is largely composed of the negative element of the right facet, with the physical and social elements of the left facet. Well-being is composed of the positive element of the right facet with the physical, social and emotional elements of the left facet, and the negative element of the right facet with the emotional element of the left facet.

9.3 Perceived modifiability

The plot also suggests a distinction between the left-hand region representing illness, and the right-hand region representing well-being, in terms of their perceived modifiability, or susceptibility to influence. Wellbeing is associated with a range of perceived influences on health over the past month, including own attitudes, behaviour, life changes, treatment and consultation, as well as expectations of long- and short-term treatment outcome. It is particularly interesting that questions dealing with the perceived severity and seriousness of the main complaint, and the length of time a patient has been attending for treatment are included in the right-hand side of the plot, where the emotional aspects are clustered.

10 THE VERTICAL DIMENSION: PATIENTS' EXPERIENCE OF MAIN COMPLAINT

Having established the main variations between the left and right sides of the plot, it is worth while considering the relationships between items along the vertical dimension of the plot.

10.1 The main complaint

The first point that emerges is that whereas the horizontal dimension deals with generalized illness and well-being, the central vertical region of the space includes all items relating more specifically to the patient's experience of the main complaint for which they are seeking treatment. This structure indicates that patients' experience of the main complaint for which they are seeking treatment does significantly overlap with general feelings of illness on one hand, and well-being on the other, but that it is conceptually distinct from either. This structure is particularly noteworthy given the heterogeneity of conditions from which patients in the current sample were suffering.

10.2 Expectations of change

At the top of the plot are a number of questions that deal with patient's expectations of change in their condition, as well as questions about perceived ability to control illness and its interference with normal life. Perhaps a particularly instructive question is the one at the top of the plot, 'How confident are you that this treatment will help?' This group of items is broadly orientated towards the future, representing the patient's own prognosis, which is shaped by the perceived modifiability or treatability of the complaint from which he or she suffers. The anticipation element of our original health mapping sentence can therefore be revised, so that it relates more directly to the perceived modifiability of the condition from which the patient is suffering.

The bottom of the plot includes items which are concerned with the antecedents of the present health status. This includes the length of time the patient has been attending for treatment, and their understanding of the possible effects of treatment. The clustering of items in this region suggests that patients interpret treatment efficacy in relation to the severity, frequency and seriousness of their main complaint.

In short, the opposite ends of this vertical dimension are marked at the top by questions concerning confidence in treatment and expectations of change, and at the bottom by questions concerning treatment history and perceived seriousness of the main complaint. We might suggest that these represent the temporal aspects of patients' experience of their main complaint, with future expectations at one extreme and past experience at the other. The main bulk of questions are centrally located, dealing with current and recent experience.

11   THE CENTRAL MICROCOSM

Having identified the extremes of the structure that define its overall shape, we are now in a position to consider what lies at the heart of patients' experiences. Those items in the middle of the plot will have the highest average correlations with other items. The central items are therefore statistically the most representative of all the items, and are likely to be the conceptual core or focus of the experiences being studied.

One item sits virtually in the middle. This is the question 'How responsible do you feel for your current state of health?' It allows of an answer that ranges from 'totally' through to 'not at all'. The central location of this item suggests that the balance of patients' experiences of health and well-being is enshrined in their view of how responsible they are for their own health. Further clarification of the significance of this central region can be gleaned from consideration of those items that are close to it.

To the left of the central region there are a set of five items, running from question 33 to question 37 which ask about the problems for which patients are seeking treatment. These are in the region of the recent and longer-term past experience of illness. The responsibility patients feel for their own state of health is therefore associated with primarily illness-related reasons for seeking treatment, which in turn are associated with past experiences of illness. More simply, seeking treatment for illness is seen by patients as an act of taking responsibility for health. But that is only one side of responsibility.

Immediately to the right of 'responsibility for health' is question 49, concerning changes in main complaint over the past month, and questions 45, 46, 47, and 48 which deal with the perceived influence of attitudes, behaviour and life changes on health over the past month. This indicates a relationship between responsibility for health and recent experiences of personal attempts to influence health. These findings provide grounds for the hypothesis that the act of seeking treatment for illness is felt as taking responsibility for health, which is preceded by experience of illness which has resisted recent attempts at control.

At the centre of the plot we have a microcosm of the total health experience. The field or background of experienced health is broadly defined by history of illness experience on one hand, well-being on the other, with specific current complaint in the centre. The central focus of this field is the feeling of responsibility for health, which is associated with the decision to seek treatment for illness, and recent experiences related to personal attempts to control health.

This structure implies that the act of seeking treatment in some sense defines the patient's current experience of health. In effect, then, there is a modification of the overall experience by the actions of the patient. These actions may be thought of in relation to the importance to the individual of obtaining relief balanced against the recognized possibilities of treatment. Thus it could be predicted that a different picture may emerge in a population of individuals who have not sought treatment, despite objectively similar health problems.

12   CONCLUSIONS: HEALTH AS AN ACTIVE PROCESS

The general structure that emerges from the above is different in emphasis from that described in the original mapping sentence. It stresses the extent to which a patient feels able, through treatment processes, to modify experiences of illness or well-being that have characteristic origins, history and possible outcomes.

This framework puts the patient in a much more dominant role than was initially conceptualized. It was proposed that the point of seeking treatment was an important one at which to examine the patient's experience of health. However, the data indicate that the act of seeking treatment plays a more important role in organizing the patient's experience of health and illness than was initially recognized.

From this analysis, health would appear to be something that is negotiated between an individual and other people whom they consult. The point of consultation appears to represent the culmination of the individual's assessment of their experience in physical and emotional terms within a temporal framework that spans long-term history, recent experiences of influence over health, and future expectations of change. The individual seeking treatment is an active agent trying to make sense of his/her circumstances and to obtain help with it. A process of negotiation results. For the patient comes with a portfolio of experiences and seeks interpretation of them, as well as actions that will alleviate their distressing components. The types of explanation given by the practitioner as well as the forms of intervention that are carried out will all help to shape the patient's understanding of what is wrong with him or her. This in turn will have consequences for the actions that the patient deems appropriate as well as his or her general state.

Such a framework has direct implications for the measurement of health. Asking a person how he or she feels at a particular moment ignores the dynamic processes that are central to that experience. Any such measurements should not only be multivariate in recognizing the various constituents that have been discussed, but also should examine the relationship between those constituents. For example, what is it that leads to a particular balance of illness and well-being or that helps to strengthen a patient's feeling of control over his or her symptoms? How does a patient's understanding of the long-term experience of illness modify their view of future likely outcomes? All these and many other questions need to be answered in an exploration of the meaning of health and the ways it might be measured.

The heuristic value of the approach in bringing together a variety of issues and perspectives has been demonstrated. But one study alone is insufficient to establish the generality of the results. Other data sets with other samples are necessary to validate the facets found here and the relationships between them. However, because it is the facets that are hypothesized, not the specific correlations between the questions, future studies could employ other questionnaires, including languages other than English. It is only by such further explorations that we can gain any confidence in answering the question of whether health can be quantitatively assessed.

ACKNOWLEDGEMENTS

We are grateful to Hana Canter for her help in carrying out the analyses on which this paper is based.

REFERENCES

Canter, D. (ed.) (1985) *Facet Theory*, New York: Springer-Verlag.
Goldberg, D. (1978) *Manual of the General Health Questionnaire*, London: NFER.
White, L., Tversky, B. and Schwartz, G. E. (eds) (1985) *Placebo: Theory, Research and Mechanisms*, New York: The Guilford Press.

TECHNICAL APPENDIX:
SMALLEST-SPACE ANALYSIS

The analysis technique of smallest-space analysis utilized in the current paper is one of a sub-set of procedures known as multidimensional scalogram analysis.

Multidimensional scalogram analysis is one of a sub-set of statistical procedures that look at the relationships between a large number of variables known, commonly, as multivariate analysis techniques. The most commonly used multivariate techniques are those that assume some underlying linearity in each of the variables and seek to relate those variables to some common latent dimensions, or factors. Factor analysis, therefore, can be seen as a procedure that looks at the correlations that every variable has with every other and seeks to establish the major dimensional components, the underlying vectors which characterize those relationships. Factor analysis, as a consequence, always leads to the identification of essentially independent dimensions that characterize the data under study. It operates on the shared variance amongst the different variables in order to develop what is in effect a list of constituents in which each plays a distinct, separate role in accounting for the way in which the variables vary in relation to each other.

Factor analysis requires the interpretation of axis or dimensions, producing models and theories that focus on the distinct dimensions which underline the processes being studied. For example, the classic personality theory identifying extraversion as one dimension and neuroticism as another is essentially a factor analytic theory.

The great strength of factor analysis is that it provides a distinct, clear framework of underlying, independent dimensions. However, it does assume that critical psychological processes are best represented as linear dimensions and it does not really facilitate the exploration of a system of interrelationships between constituent factors.

The cluster analysis approach to multivariate statistics is in contrast to the dimensional approach of factor analysis. In cluster analysis, each of the variables is assigned to one cluster. There are some more sophisticated procedures in which multiple cluster membership is possible, but that quickly becomes confusing except in very special circumstances. The basic approach of cluster analysis is to look at the relationships between variables and decide whether or not the relationships within a particular sub-group of variables is higher on average than any one of those variables' relationships to other variables outside of the sub-group. In other words, a criterion is established to determine whether or not the relationships between a group of variables is high enough to define them as a cluster distinct from all the other variables. The criterion therefore, determines how many clusters might exist. A very stringent criterion would mean that you had a lot of clusters with few variables in each, whereas a very relaxed criterion could imply that virtually all the variables are related to each other in one global cluster. The arbitrariness of the criterion therefore gives rise to the fact that most cluster analysis is, in effect, a hierarchical procedure, in which clusters nest within clusters on the basis of their particular criterion being applied to them. Conceptually the dominant scheme behind cluster analysis is whether the variables being examined do or do not have properties that make them related to the other items within the cluster.

If there was a strong theoretical basis for assigning items to clusters and for determining the meaning of the hierarchy, then this procedure can be very powerful. For example, the most significant cluster analysis framework is the Darwinian model of speciation. Plant and animals form two distinct clusters on the basis of clear criteria. Animals can be divided up into groups, say on the basis of whether they are invertebrate or vertebrate and whether they have exo-skeletons or endo-skeletons and so on and so on. This hierarchical cluster system provides the classification of species. It is scientifically fruitful because each of the steps of the hierarchy have some theoretical underpinning to it. However, in areas of behavioural research, in which there is no such basis for determining the criterion for assigning entities to groups, cluster analysis can prove quite arbitrary.

The multidimensional scaling approach in a sense sits between the clustering and the dimensional approach. In this framework, the distances between variables in the multidimensional space represent the dissimilarities between the variables in terms of their correlations or associations. The spatial configurations that are produced can be interpreted as dimensional structures, interpreting the horizontal and vertical axis, or as clusters by identifying groups within the multidimensional space. As will be elaborated below, the facet approach looks for regional structures rather than dimensions or clusters.

There are many algorithms available for representing correlations as distances in a multidimensional space. The smallest-space analysis procedure, developed by Guttman and Lingoes is unusual in only seeking to represent the rank order of the association matrix as closely as possible to the rank order of the distances in the multidimensional space. Mathematical examination has demonstrated that by working with the correspondence between the two rank orders, a better fit can be established in a small dimensionality than if the absolute values are used. This is why the particular procedure is called smallest-space analysis.

In all multi-dimensional scalogram procedures there is a question of the number of dimensions that it is appropriate to extract. However, all such numerical procedures suffer from the fact that they have no substantive theoretical base and are little more than rules of thumb to indicate when there is a reasonable degree of relationship between the multivariate summary and the original pattern of correlations.

The facet approach introduces a rather different perspective on these analysis systems. It does not start from the view that the statistical summaries do of themselves reveal objectively valid aspects of reality. Rather, it works from the constructivist perspective that the scientist's task is to provide a logically convincing description of reality. There is therefore always a search for a correspondence between some conceptual model and some distinct aspect of the structure of the observations.

Facet theorists look for ways of classifying the observable phenomenon in theoretically fruitful and interesting ways. They then seek to verify that these modes of classification are reflected in the empirical structure of the observations. It is accepted within the facet framework that one empirical structure may be open to a number of different facet interpretations. The interpretation that is accepted is the one that is most fruitful and logically coherent.

Another important aspect of the facet perspective is that any observations can be looked at from the point of view of a number of different facets. For example, the question 'How ill have you felt over the last few days?' is a question that contains facets that refer to the respondent 'You'. It is feasible that some other person's view of your illness could have been the basis of the question, so that a question such as 'How ill does your doctor think you have been over the last few days?' would contain a different element of the person facet. Another facet present is the one that refers to 'feeling ill'. The question could have been phrased in terms 'How difficult has it been for you to move around?' or of a more direct symptom such as 'How high has your fever run'? A third facet that is present is the temporal one in terms of questions about the last few days.

This one question, therefore, could play a role in a number of different facets. Empirical research is necessary to determine which of the possible facets do actually exist in the observations and what might be the relationships between them. So, although it may seem self-evident that all the questions containing reference to the 'past few days' should correlate highly with each other, that would only be the case if the issues relating to the past few days were experientially correlated as well as conceptually. Questions could readily be developed which would not give high correlations simply because they included the phrase 'the last few days'. For example, questions like 'Have you thought a lot about your childhood over the last few days?' is not necessarily going to correlate highly with the question 'Have you felt ill over the last few days?' This is an important point because published material within the facet framework tends to emphasize structures that have been found to have empirical support. Those analytical structures tend to go unreported. There is, therefore, an apparent indication in the literature that facets that underlie the structure of questions are always established. This is certainly not the case.

The conceptual clarity of the elements of a facet are therefore as much a part of the validation of the facet as their observed closeness. It follows that if items are given a distinct definition the evidence for them will be established by finding regions that are clearly related to that facet element. There will always be the possibility that some items are on the borderline between regions because they have some elements in common with those items within the region and they have elements in common with items in the adjacent region. Furthermore, because every item has a number of different facets that make it up there will always be the possibility of a multidimensional structure in which facets overlay each other in more than two dimensions. It therefore follows that the drawing of regional lines on smallest-space analysis is an indication of the distinction between elements. Currently these lines are drawn by the researcher on the basis of the meaning of the items. However, computer programs do exist that will draw the lines, provided the faceted definition of the items is already established.

The facet approach, therefore, can be used as a hypothesis-generating, or exploratory, procedure whereby the spatial structure is used to suggest the underlying facets. Such hypotheses can then be tested in subsequent studies by seeing whether or not such regions do indeed occur with other sets of data. If the facets themselves can be drawn from existing literature then the approach can be used directly to test facet hypotheses.

In either the hypothesis-testing or the exploratory mode the number of dimensions that are chosen for interpretive purposes are influenced as much by the interpretability of any given dimensional structure as by numerical guidelines that come from the measures of fit between the correlation matrix and the distances in the space. Of course, if there is a very low degree of fit then models are liable to be unstable, but this simply refers back to the need to replicate any findings rather than to build strong theoretical structures on the basis of only one set of data.

The great advantage of the facet approach is that it allows for the unfolding elaboration of conceptualizations. Broad imprecise facets can be starting-points which are elaborated as further data and conceptual clarity evolves. The approach also has the value of providing a system of relationships between facets rather than an independent dimensional structure as in factor analysis or the inevitably hierarchical structure of cluster analysis.

In those domains in which there are a wide range of conceptualizations and a number of varying perspectives, the facet approach has proved particularly fruitful in drawing various strands together. It provides a systematic framework open to test and elaboration whilst not demanding very high levels of measurement.

The main pitfalls of the facet approach are around the demands it places on the conceptualization and definition of the central facets. If these are not couched clearly and do not have an overt and logical relationship to the questions being asked, then researchers may convince themselves that they have identified a facet structure when their results do not really support their claims. It is, therefore, essential in reading and publishing facet studies that it is possible to cross-refer directly to the questions and the points in the plot. This way the experienced reader can establish for himself whether or not the content categories to which the questions are assigned do have some logical coherence to them and that the regional structure makes scientific sense.

REFERENCES

Canter, D. (ed.) (1985) *Facet Theory*, New York: Springer-Verlag.

Everitt, B. S. (1977) 'Cluster analysis and miscellaneous techniques', in A. E. Maxwell, *Multivariate Analysis in Behavioural Research*, London: Chapman & Hall.

Tabachnick, B. G. and Fidell, L. S. (1983) *Using Multivariate Statistics*, New York: Harper & Row.

14

REVEALING THE SELF
Biographical research as a method of enquiry into health

Jo Lebeer

1 TELL ME THE STORY OF YOUR LIFE

Jacques Lusseyran became totally blind after an accident when he was eight years old. Nevertheless he developed a special sense of 'total' perception, including his intact sensory organs and his intuition. In this way, during the Second World War, he became one of the leaders of the French Resistance where he was responsible for editing an underground newspaper. After three years he was betrayed and sent to a concentration camp. There he survived pneumonia, starvation, torture and typhoid fever. He describes his life history in his book *And There was Light* (Lusseyran 1965), in which he writes:

> When you asked me: 'Tell me the story of your life', at first I didn't like it. But when you added: 'I would like to hear why you love life so much', I started to love the idea, because that's the good subject.

This example makes us wonder how it is possible that this man had such strength that he survived very severe illness in extremely adverse circumstances and without any medical help. His answer gives us a clue pointing in the direction of life history: to understand health we need to understand a man's life history and his most central existential experience. Only in this way could we gain an answer to the question: what makes a man love life ? It is easy to love life when nothing happens or when circumstances are favourable. But to be able to love life intensely despite the most extreme physical and psychic traumatic experiences, not only elicits our human admiration, but also our scientific curiosity. Lusseyran's book questions current belief systems about the relationship between health and risk factors which are at the basis of many current health promotion campaigns and health research. General assumptions are that health is largely determined by *circumstances* like good genes, absence of all sorts of risk factors , a 'healthy' life-style, a balanced diet, etc. Reading Lusseyran's life history, however, makes clear that his health cannot be explained by positive endogenous or

205

exogenous conditions. To understand health one has to go deeper and look at an inner level: the level of emotional, cognitive, and existential experience, which is one of the domains reflected in a person's life history.

In this chapter I want to show the necessity of using biographic studies for health research and to outline some methodological problems and possible solutions. My purpose is to delineate a framework for using new instruments in order to gather and interpret biographical data. The prime reason for our methodological interest is our conviction that sound scientific discussions can only be made possible when the methodology is clear. Our hope is that this will reinforce scientific research to deepen the mystery of the mind–body relationship.

2   THE BACKGROUND TO BIOGRAPHICAL RESEARCH

Interest in the relationship between biographies and health is very old and widespread: it can be found in all historical periods and in different cultures, for example in yoga, in Hippocrates' teachings, or in the culture of native Americans. But even in the medical history of the West, which has largely devoted its endeavours to technological developments in fighting disease, an interest in the human aspects of health and disease has always existed.

The famous British surgeon Sir William Osler wrote: 'It is more important to know what kind of man a disease has, than what kind of disease a man has' (Osler 1921). Her Majesty's obstetrician of last century, Sir James Paget stated that 'one has to cure a woman's depression first in order to cure her breast cancer' (Paget 1870). How many people nowadays would know that Bismarck's personal doctor, the Head of Berlin's University Hospital Ernst Schwenninger, was an avant-garde holist who drew no distinction between so-called somatic diseases and psychologically caused illness, but who saw every illness in the context of a person's whole life (Groddeck 1930)?

Freud's psychoanalysis created a paradigm that beneath the conscious life experiences lies an invisible world of vivid darkness in the unconscious human life history. Freud attributed disturbance of behaviour to such unconscious repressed experiences. His disciple, the German psychoanalyst Georg Groddeck, practised psychoanalysis – with amazing success – on what are usually considered to be 'somatic' patients, people with cancer, hyperthyroidism, gout, etc. (Groddeck 1983)

Another pioneer in the understanding of how disease arises out of the life history, was Victor von Weizsäcker, professor of neurology and psychiatry at Heidelberg University in the 1920s. He investigated the biographies of sick people, and developed his theory of *pathosophy*, literally the 'wisdom about the origin of human suffering'. In his opinion disease occurs in the course of life whenever something essential is not fulfilled. 'Disease is the manifestation of the un-lived (*das Ungelebte*)', he writes, 'it is a way of dealing with the world . . . a transformation of mind into matter'. According to von Weizsacker it is essential to investigate the unconscious realm of an individual to explain health and disease. And he observed that people become healthier when they make important choices in their lives: '. . . das Beste tun wir nun wenn wir müssen, und dieses Muss kommt denn gewöhnlich in dem Kleide des Schlimmsten'. Freely translated it would sound like: 'we only choose for the best when we seem to have no choice, and this situation usually appears to us as the worst' (von Weizsäcker 1957). Von Weizsäcker sees disease as inviting, almost forcing, someone to make life-changing choices.

Recently the sociologist Aaron Antonovsky, professor at Beer-Sheva University in Israel, has emerged as one of the leading researchers of biographies in relation to health. Antonovsky investigated people who have stayed particularly healthy (which is not the same as never having been ill) in spite of massive doses of stress such as living in a concentration camp. He concludes that those people stay healthy who have developed a strong *sense of coherence*, which he defines as existing of three components: the ability to convey *meaning* to life circumstances; the *manageability*, i.e. the experience that one can cope with life whatever happens and one will be able to find solutions; and the *comprehensibility* of life, i.e. the experience that life can be understood. Antonovsky, however, has no explanation of how and why some people have a strong sense of coherence (Antonovsky 1987a).

Psychology and sociology may be very familiar with the context-bound roots of health and disease, but the medical world seems to have largely ignored it during this century. Subjective experiences of patients have long been kept off-stage in scientific discussions. However, this is changing. Some of the leading medical journals have published individual reports of patients that recovered from chronic, incurable diseases (Fiore 1979; Cousins 1976). These reports are not only interesting from a human perspective, but they challenge fundamental medical belief systems and create confusion. For example Norman Cousins described his recovery from ankylosing spondylitis, an inflammation of spine connective tissue, leading to progressive stiffening and immobility. Instead of resigning himself to such a pessimistic ordeal he wondered why he had become susceptible to a degenerative disease, and said to himself 'If negative emotions produce negative chemical changes in the body, wouldn't positive emotions produce positive chemical changes?' He decided to stop the prescribed medication and composed his own programme of recovery, consisting of whole-food nutrition and vitamin C in very high dosages. After the acute phase he went to a hotel, because he was convinced 'a hospital was no place for a person who was seriously ill'. He used comic movies to strengthen his optimism and to make him laugh again and discovered that this not only reduced his pain but also his blood sedimentation rate. His new attitude towards life, his will to survive and to be happy were very important in the recovery from his illness. Cousins' article had a widespread impact: it generated a stream of 3,000 letters to the editor (Cousins 1979).

The amount of literature, medical as well as literary, in which people write personal accounts about their lives and the recovery from chronic illness is growing vastly. Clearly a study of life histories of such 'healthy sick people' and 'healthy healthy people' can be very relevant (Lafaille and Lebeer 1991).

3 FIRST OBSERVATIONS ON LIFE HISTORIES AND HEALTH

In the past fifteen years I have studied the life histories of hundreds of people who have gone through a healing process, first as a family physician, later as a counsellor, and recently in a more systematic way as a university researcher. Family physicians are in a privileged position in the health-care system to get to know the context of a person's disease and health, since they see patients in all stages of life and with minor and major ailments (Huygen 1982). It was as a family practitioner that I became interested in people's lives. I started to ask them what happened, when, and why? How do they look at life, at the future, at the past?

I am well aware that I cannot present statistically valid numbers here. Moreover I can only give data from a population of one practice in an industrialized urban area with a reasonable standard of living; this inevitably limits the generalization of observations and interpretations. However, a common pattern returned so often that a relationship between biography and health is beyond doubt.

It became apparent that acute illness such as pneumonia is very often linked to biographical events. I also worked with people suffering from a variety of long-lasting illness or chronic diseases such as essential hypertension, cancer, rheumatoid arthritis, asthma, eczemas, allergies, depressions, brain damage, etc. There is a great variety in how people cope with illness and with life, and an equally large variety in the patterns of a disease. Grossly, three categories can be distinguished.

Sometimes there is a cure, even in so-called incurable disease states, but this is unusual. From the point of view of medical science this is the most interesting minority to investigate, because it can clarify healing mechanisms.

The second category consists of a large number of people who do not get better, not in a biological or psychological sense. For many people disease is meaningless and is only associated with loss. They are usually the ones that do not see how they can change their living conditions and who let themselves be governed by the flow of life with sometimes very adverse consequences.

On the other hand there is a third group who have found creative ways of dealing with their disease. Although the desire for complete biological recovery was for most people a primary motivation to examine and change their life-style, in many instances this goal was only partially achieved. Patients in this group often realized that other values were more important to strive for, and that they did not need to be absolutely free of symptoms to lead a meaningful life. And paradoxically, when fundamental choices were made towards 'life' – more freedom, autonomy, (re)finding a sense of meaning, purpose and a source of joy in life, acceptance of life pain – in many cases the result was also more biological health or even recovery. But in any case they experienced themselves as more healthy even without having a healthy body.

Here we have a first indication of what 'The Self' may be like, which we will explain further below. We have to keep in mind, that here too, as in other chapters of this book, we are using 'health' in the old etymological sense here: health as a synonym of wholeness, i.e. body–mind vitality.

4   A SYNTHESIS BETWEEN INTUITIVE KNOWLEDGE AND SYSTEMATIC ENQUIRY

Biographical research is the understanding of life's story. Therefore an obvious question which is of interest to medical science is how to lift the relationship between biographical research of health beyond a level of impressionism and intuitive knowledge. Even if almost all major scientific discoveries have been the product of much intuitive imagination, the task of science is to check hypotheses with new data. A sound science requires a maximum of transparence and verifiability. Can then research into life histories be made more transparent?

A subject's sense of coherence can be researched by means of questionnaires, as Antonovsky has done. However, to research processes which have to do with consciousness, the 'unconscious', the mind, and emotional experiences, is much more difficult. There is no reason why these processes should *a priori* be exempt from scientific observation, but they require caution. Freud has often been criticized for not being very scientific: he used much *hineininterpretierungen* (circular reasoning), and there was no possibility for other researchers to check his data (Spence 1982). In Freud's days, however, the only way of registering data was to write them down. Today tape recordings, video and computer have widely increased the possibilities of registration and interpretation.

5   SOME DEFINITIONS

When studying a biography we study a complex interaction of factors in relation to different contexts in which a person lives or has lived: social environment, work relationships, ecological environment, family, etc. as well as the unique, inner, psychological and transpersonal development.

209

A distinction has to be made between levels of biographic investigation. There is an *outer* level of life events, actually observable by someone else. This includes, e.g. a person's case history (i.e. that part of a person's biography which is usually known to the doctor), and also his life history (seen as a sequence of events, which is often not known to doctors). This can be called the *outer biography*. But there is also an *inner biography*: how someone experiences his life, including how life is experienced unconsciously, in individuals or societies. This is the difficult part to research. The inner biography is not readily accessible for the rational left-hemispheric mind, but may be expressed in terms of right-brain hemisphere communication: art, imagination, fantasy, myths and symbolism.

Many scientific discussions which do not make this distinction are unproductive. One example is the question whether or not the cause and course of cancer can be psychologically influenced (Spiegel 1989; Grossarth-Maticek *et al.* 1985). The Holmes and Rahe scale of stressful life events is often used to look for a link between psychological causes and illness. When no correlation is observed it is declared that there is no relationship (Andrews and Hall 1990). But in this case the inner biography may have been overlooked. An apparently 'uneventful' life history can cover a profound sense of victimization. Or vice versa, a life full of adverse situations and unfortunate circumstances can nevertheless lead to a very mature personality, who has a strong sense of being in control of his life. Michael Lerner, for example, interviewed people who recovered from cancer regardless of the sort of therapy they underwent; he was interested in what elements in their biographies might have contributed to their health. He found some recovered patients with nerve-racking lives, but he also found some with uneventful lives (Lerner 1985). This inner/outer distinction was also demonstrated by Steven Locke, who measured activity of natural killer cells (NKC) in a student population in relation to the life-change stress-ratings. When only objective variables were looked at no significant correlation was found. However, when the students were asked how they had experienced their crisis, those who had a higher sense of hope and purpose showed a significantly higher NKC activity (Locke 1984).

6   WHAT IS 'REVEALING THE SELF'?

What is the Self? A few concrete examples of biographies[1] will make it easier to understand the definition.

Example 1

Mrs K. describes how she recovered from anorexia when she was young. She was in hospital for six months close to death. Talks with a psychotherapist did not help. One day a friend's visit made her aware that she was in fact choosing to die, and this was the turning point: a sudden shock-like awakening of consciousness. In this fragment of an interview taken twenty-five years later, she describes what happened:

> I wanted to escape from my environment as soon as possible, to lead my own life, a new life, to start my life over.
>
> At last we got married, and that was my liberation. Finally I was the master of my own life, I could arrange things as I wanted. From then on I always wanted to live and be aware of what I was doing: in my relationship with Michael, with my children, the clothing, many things. I experienced this as a time full of dynamism, I was very dynamic in taking up every challenge. The anorexia slowly disappeared.

At about the age of forty Mrs K. developed essential hypertension. Strikingly her initial wish to lower blood pressure with homeopathy and diet did not materialize. After a year her pressure was dramatically high. Then she accepted anti-hypertensive drugs. In this fragment she describes the process of self-enquiry and her inner development, leading eventually to a lowering of blood pressure, enabling her to stop taking pills.

> So I started to ask myself: what's this high blood pressure signify?
> *Are these questions important to you?*
> Yes, indeed, I began to wonder, after the worst fears were over. I was taking those pills of the GP, I was feeling a bit better, the blood pressure was falling down, I was calming down. Yet I wanted to know where it all came from, because I don't want to go on for ever taking pills, but he [the doctor] couldn't . . . he says 'it's so complex' . . .
> *Did you have any idea yourself?*
> . . . well, you know, my first thought was: it's my way of life, much too busy, too many things at a time, yes, I used to think that for a while. Yet by reading about it, and by starting to look at myself, I realized, it's got something to do with me, . . . It's not the circumstances, but it's me, the way I'm dealing with everything.
> *And how did you think you were dealing with everything?*
> . . . I thought I was running past myself, that I didn't take care of myself enough . . . at that time I didn't notice that there was a huge anxiety under everything I did, so, when I got involved in some kind of activity, or when I was organizing something, I always felt very anxious to make a good performance.
> *When did you begin to realize this?*
> About three years after the onset of the hypertension, I realized that it all came down to my anxiety, and this I linked to my fear of my father, which I had also in front of everybody.

211

Nobody has told you this, have you found out yourself?

No, I've found out myself. Maybe, yes, the most important influence came from reappraisal-counselling, that's where the turning point was, and the beginning insight in where it all came from. That's where I started to work on my fear of my father, and I realized that on a more fundamental level, it had to do with fear of meeting other people, and with security, indeed, I'm simply scared to death of other people, and it's when I'm close to other people that my blood pressure goes up, it's fear of being rejected, fear of not doing well enough. Yes indeed, that was what it was . . .

. . . that was the way we tried to come across our emotions, to become aware of what sort of pain, or what sort of joy, what's really there at this moment, . . . I've learned to be angry there, I couldn't be angry before, never could be angry with my father, I killed my father there, but I've also learned to appreciate him, because of all the good things he has done to us, because that's very important, you know, to know the difference his patterns in behaviour towards me, and who he really was . . . I know in my first group meeting, I started to weep, to cry, I shouted my anxiety out of myself, my fear of my blood pressure, my fear of dying, thus, . . . that's been a big liberation, to be allowed to become angry, all your feelings, everything, everything, just spit it out, yes that was really very important!

Then the second step is to re-appreciate yourself, to let go of the rejections of the past like 'you're not good enough, you're not decent'; to realize 'I am good enough, with all my mistakes, and all my bad habits', and to experience this from other people in the group, this esteem for who you really are, that's what gives back security feelings, that's what restores the belief in oneself, and when you have that, the whole world may go on to disapprove of you, but you stop caring . . .

. . . *And how is the relationship with your father now, has there been any change?*

At last my anger has gone, it's taken a couple of years, I had to fight for it, to scream, to beat, to let all out. . . . but that has caused me to look at my father better, I started to look beyond his rejection, why he was behaving like that, why he was who he was . . . so I gradually started to feel something like forgiveness: where the anger, the hatred dwelled, now there was a place for forgiveness. 'I forgive you for what you have done to me.' That was extremely important, but you can only get there, when you have let go of your anger, only then you are able to forgive.

Do you experience this as a spontaneous process, or do you experience it as a choice? How does it really happen?

I guess one makes a choice for it, to look at him, not from your anger . . .

Shortly after this interview her father died. He had stayed in her home for a few days, an experience which she described as completely peaceful for the first time in her life, towards her father.

Example 2

Mrs V. developed breast carcinoma at the age of 41. She was very determined that she did not want an amputation or any chemotherapy afterwards. Instead, after just local surgical removal of the breast lump, she first started with diet and immunostimulating drugs, derived from *Viscum album* extracts. At the same time she started to reflect on her life. Her biography was on the surface very normal. She was the eldest of four siblings; nothing dramatic had really happened; she became a nurse, married, got three children, has had no financial problems. Nevertheless, she admits that a few years before the onset of her cancer, she was vaguely depressed. Although she described her marriage as fairly happy, she was missing something. She was very devoted, always working, giving her life to husband and children, suffering from a dominant and criticizing mother, having no psychologically nurturing contacts. She gradually discovered that she was lacking inner autonomy and the experience of being loved and of loving. Through a process with lots of crises, allowing herself to look at her fears, anger and sadness, she came to the experience that she could be independent of her 'inner criticizing mother'. She became more free of her past experiences and belief systems which had caused so much inner unhappiness. She also discovered her power of love. Ten years after the diagnosis of breast cancer, she is still free of metastatic spreading.

These two case and life histories evoke several questions. A first group of questions concerns aspects of validation and causality: has there really been a cure, and if so, how can it be explained? The conditions described in both examples are generally known to be 'incurable', or at least very hard to cure. The good outcome reported in the above examples questions such deterministic concepts. But we must not jump too soon to conclusions. Thinking within the classical causal paradigm, criticisms can be made that the length of observation has been too short to allow the use of the word 'cure', in the sense of 'absence of disease'. It is not at all certain that there will be no long-term damage from hypertension in the first case and in the second example a ten-year-long absence of metastatic spreading is not a guarantee of absence of cancer. Another criticism is that the number of cases is too small to allow conclusions, and a third possible criticism is that a good outcome could be due to sheer coincidence.

On the other hand a good outcome is not necessarily identical to cure. It is a combination of increased sense of physical wholeness with increased meaningfulness, joy of living, and freedom. Defined in this way, the women

213

in both examples showed beyond any doubt a significant improvement of health.

A second group of questions concerns the nature of health and healing process, which Antonovsky has termed *salutogenesis* (Antonovsky 1987b). Looking at the examples, it is not clear what kind of intervention could be seen as mainly responsible for salutogenesis. Various interventions have been made on various levels. On the surface, it looks as if specific 'material' substances or acts are the cause of health: in case 1 anti-hypertensive drugs have been taken, and in case 2 an operation has been performed. It is clear that both have been extremely valuable in the acute phase of the disease. They can be named *disease-eliminating procedures*. Once the acute phase has passed these interventions have not been sufficient. Different changes in life-style have been made, all aiming at restoring 'resistance to disease': change of nutrition, reduction of stress, e.g. relaxation techniques, and reorganization of time and work load. This category can be named *healthy-physiology-promoting procedures*. The use of *Viscum album* extracts as adjuvants in cancer therapy, which has been well documented (Leroi 1977), can also be seen as belonging to this category, as can a range of other medicines, methods and techniques known as 'natural or holistic medicine'.

However, to look *only* at an outer level to explain health is not enough. In both cases health continued to improve even when a number of 'outer' measures had been stopped. In case 1 anti-hypertensive drugs had been stopped, and nevertheless the blood pressure did not rise, and there were other periods when Mrs K.'s blood pressure rose again dramatically even when she was taking sufficient doses of beta-blockers and diuretics. In case 2 health continued to improve (and cancer did not return) even after minimal surgical intervention, which is considered a highly insufficient treatment for breast cancer. The same objection applies to making a single causal link between 'category two procedures' – Viscum album injections and diet – and absence of tumour spreading, because injections were stopped and a strict vegetarian diet was abandoned.

In the above examples salutogenesis (after the initial stage) was primarily independent of therapeutic procedures, and changes in social context. Social conditions changed as a result of choices made by the protagonists. There must be a hidden variable.

My hypothesis is that the *hidden variable* must be looked for at an *inner level*. This is less tangible, it is not a matter of clear-cut procedures. It is also deeper than psychological or psychosomatic information. Clearly the reader will be able to sense, to experience, that the women in the examples have developed something special, a special energy, a quality of being. It is obvious in the text, and it is even more obvious when meeting somebody like that. They have gone through an inner process leading toward more autonomy, inner strength, forgiveness, creativity. A more efficient handling of painful and stressing emotions, reducing physical stress, and life-style changes are but a result of this process. This inner process can be described as the unfolding of the Self.

So what is the Self? It is not only difficult to research and operationalize, it is difficult to experience, and it is difficult to express in words. The Italian psychiatrist Roberto Assagioli postulated a difference of order between the Ego and the Self, thereby going beyond Freud's theories (Assagioli 1965). The Ego can be considered as the personality: the form in which we are observable as a person to other human beings. The Ego consists of a vast realm of conscious as well as unconscious experiences. The Self, on the other hand is the inner core, the essence of being human. Values, meaningfulness, transcendental (existential) experience, love, wisdom, intuition all belong to the order of the Self. Carl Gustav Jung, as well as many psychologists of the humanistic psychology tradition, used the same distinction.

7 OBSERVING THE SELF

We will focus now on the problem of how to get access to the Self. We will discuss here the possibilities of *revealing* the information from the inner level of the Self in a scientific way. This is the first phase of data gathering in a scientific investigation. These data are essentially of a qualitative nature. The next phases of the scientific process – data analysis, interpretation, drawing conclusions, verification and control, and building causal networks – will not be discussed here. This has been done extensively elsewhere (Miles and Huberman 1984; Glaser and Strauss 1967).

The Self as such cannot be observed, but its effects can. It becomes visible through subjective experience, belief systems, coping behaviour, choices. A first access is a direct interview or a series of interviews, which specifically enquires about these areas. Relevant questions could be:

- What happened in the environment (family, work, school) prior to onset of disease?
- How did the person cope with crisis, loss, disease? How long did he/she feel victimized in difficult situations? How much creativity did he/she develop to look for solutions?
- What kind of transitions occurred in a person's life, and how did this person cope with them?
- How did the person deal with his/her own emotions and those of someone else (anger, pain, joy)?
- What important choices did this person make in his/her life?
- Are these choices made out of extrinsic motivation (= 'because somebody told me') or intrinsic motivation (= autonomous decisions, 'because I want it')?
- How are these choices experienced? (Meaningful, imprisoning, liberating, weary?)

- Does this person take the important (life-changing) choices out of intuition ('I do it because I deeply feel it is right to do so')?
- Are the wounds of the past really healed, or do they still interfere with life now?
- Does he/she experience his/her own life as meaningful generally? Does he/she look at life in general as something meaningful?
- Does he/she experience closeness / to be loved by someone / love toward someone? How has this been achieved? What was the process?
- Does he/she experience commitment to somebody or to some cause?
- Has he/she ever had transcendental experiences?
- What are a person's resources for coping with life?

The first condition of collecting relevant data about the subjective experiences of subjects is to build a good relationship between interviewer and interviewee. An interview is a meeting between two human beings who cannot but become affected by each other. For example, when interviewing Mrs K. I had the impression that probing again into this anorexia-episode and into the anxiety about the hypertension, and all the things that had to do with it, was somehow hurting her again, like breaking the scab of a wound. This feeling was confirmed by taking her blood pressure in the middle of the interview: it was 16/11. Nevertheless afterwards she told me that she did not mind going into the wounds again, because that was a check for herself to see if she was really healed. Bringing her in contact with her pain and anger brought her also in contact with her capacity for love. Strikingly, the blood pressure after the interview was 14/9!

The close connection between interviewer and interviewee is supported by research done by James Lynch on hypertension (Lynch 1982). He used a semi-continuous computerized monitor of blood pressure which reacts very sensitively to emotions. Lynch warns against inquiring about painful life experiences with someone who has essential hypertension, as this can raise blood pressure dramatically. The interviewer influences not only what the interviewee says, but also his body! In other words: by gathering data, one changes them.

This is compatible with the new scientific paradigm emerging which has been formulated in its essence by Nobel prize winner Ilya Prigogine: 'Science has to cease to be a monologue, it has to be a dialogue with nature' (Prigogine and Stengers 1984). Prigogine developed his ideas from chemistry. Because there has been a tendency in health sciences and in psychology to accept as truthful only information which can be expressed in quantities, information belonging to the level of the Self has been banned from scientific discussion. It is doubtful whether an inner process like the development of the Self can be observed by using standardized questionnaires. A questionnaire is a monologue. Biographic research is essentially a dialogue with a human being. Only in a dialogue, can interviewees' experiences of pain, anger, love and forgiveness emerge, provided the interviewer (doctor, therapist) is open to his own Self. To see the essence of a healing process one has to be a participating observer.

People are sometimes afraid to talk about their experiences, certainly the most painful ones, but also sometimes the pleasant ones (such as a secret affair), or even transcendental ones. Sabom, a cardiologist who became interested in the experiences of people who recovered after a clinical death, wrote that almost all people who had a near-death experience were afraid to talk about it because other people might think they were going crazy (Sabom 1982). If people feel that their experiences will be dismissed, judged badly, laughed at, or treated with disrespect, they will remain silent. On the other hand, if the interviewer is open and respectful toward the experiences of the interviewee it is very likely that relevant material will emerge.

A second way to reveal the Self is to make use of symbolic representations within a biography. Many authors have described how pictorial methods throw light on the inner development (the inner biography). The inner biography is a reflection of how people learn to cope with events and circumstances they encounter. This can best be illustrated by an example.

Example 3

A woman of 27 who in her medical history had had persistent urticaria and Quincke's oedema, for years did not dare to go out without anti-histaminic medication. She recovered. In the course of ten interviews it became clear that the allergic episodes had been triggered by meetings with her alcoholic father, whom she resented deeply. When the father got cancer of the oesophagus, which resisted all therapy, the woman remembered the fairy tale of the princess and the frog, which at one of the interviews, she had advanced as the symbol which most accurately represented her life-long difficult relationship with her father. This story helped her to overcome her resentment towards her father, whereupon the allergy problem was overcome.

One way to outline a symbolic biography is to ask someone to look for a fairy tale which suits their life history best. Another way is by doing an exercise where one is requested to visualize one's life as the journey of a hero, who encounters different situations. This might sound strange and unscientific, but, provided all interpretations are discussed and checked against other sources, it can yield very useful information.

There are snakes in the grass, however. When someone is being asked 'How did you feel about it?', and the answer is 'Fine!', this is not necessarily true. Verbal information can so easily mislead. But the body does not lie. Hence, a third way to reveal the Self is to record non-verbal information. Neuro-linguistic programming includes a detailed system to encode non-verbal information, which is used to get access to the structure of subjective experience (Tilkin-Franssens 1991). Discord between verbal and non-verbal (body) language is especially likely to occur when the interview touches upon painful experiences. Almost every biographic interview which is related to health at some point comes across the wounds of the past. It requires tact and care from the interviewer, as well as attention to watch for non-verbal utterances. For example the interviewer can say: 'Do I notice rightly that you get somewhat tense here?' (that you blink your eyes a lot, hold your breath, look away, cross your feet, etc.). Or he can communicate his own feelings or body reactions, like ' I notice that I am holding my breath here', inviting the other to react. Non-verbal expressions are highly individual, and do not allow a cookbook interpretation.

8 HOW TO BRING SCIENTIFIC METHODOLOGY INTO BIOGRAPHIC RESEARCH

From the explanation above it is clear that the quality of data in biographic research depends a great deal on the quality of the interview. This is a scientific paradox. If observation also depends on subjective experience a criticism can be made that biographical research such as we propose is falling back into esoteric science. However, this is not true. Every scientific discipline uses instruments which demand skill and time to learn to use. One cannot learn to handle an electron microscope, to interpret cervical smears, or to read an ultrasound scan just from reading a booklet: it also needs critical self-observation, supervision and practice. One of the instruments used in biographical research is awareness of subjective experience. This also requires practice and training. This is often overlooked in social, medical and psychological sciences.

Because biographical research is analogous to historiography, criteria of validity adopted in the methodology of historical research (Ricoeur 1955; Vansina 1972) can be applied to biographies for health research. A first criterion is the competence of researcher. It is the issue of whether the researcher is competent to discover the truth. The specific competence needed for a biographical observer lies in his capacity to relate as a human being, to use intuition in an appropriate way, to experience his own subjectivity, to observe expression in body language and to use symbolic language and expression.

Some essential qualities of a good interviewer are:

- *Open-minded/open consciousness*. This is an attitude which is even communicated without speaking a word, e.g. by eye contact, or the manner of greeting. It means to treat all experiences and expressions of the other with respect, whatever the strangeness or the content may be. One cannot 'play' it, it is not just a matter of a good behaviour to be adopted by the interviewer. It is felt on a deeper level by the interviewee.

- *Non-judgemental attitude.*
- *Inner development.* What is the life experience of the interviewer? It is rather difficult for an interviewer to actually observe transcendental experiences if he never has had any such thing before, ignores them, or worse, even denies them. The same holds for qualities like 'compassionate love'. Can the interviewer make a distinction between superficial love (sympathy, friendliness) and true love, which has a totally different quality? The only way to perceive it is to experience it. Most people have had such meaningful experiences (e.g. 'love from the heart', 'unconditional love', 'feeling one with the whole', 'deep joy'), but our culture and certainly science has a tendency to rationalize them away. The interviewer has to be able to acknowledge them. Similarly, there is a fundamental difference between 'conciliatory behaviour' and 'true forgiveness'. One cannot perceive the difference by counting the number of laughs or observing a specific element of behaviour. For example, a person who has an attitude of autonomy, who dares to say no, who is independent, can have an experience of forgiveness after a long-lasting conflict without apparently changing behaviour. Only an experienced observer will notice the difference of quality.

The second criterion is exactitude. This is the question of how closely observations match reality. A researcher should keep the following questions constantly in mind:

- How do I know this? Where do I get this information from? Is this my observation, or is this an interpretation?
- Has the person I have interviewed a good capacity for self-observation? Was he/she able to know what happened?
- Were the circumstances of the interview optimal?
- Has the person any reason to lie, or to conceal the truth? Are the sources I use reliable? What other sources do I need? Did the interviewee have enough sense of accuracy to recall the past?
- Are there things which I never perceive? (e.g. do I have difficulties in observing someone else's inner pain, or anger?) Or are there things which I always see in the same way? (e.g. do I see aggression every time the person is merely nervous?)

It is impossible to uncover the whole truth of someone's life history. One inevitably makes selections. The researcher therefore has to make explicit which elements he wants to control. With a video more information is obtained about the visible non-verbal behaviour. On the other hand some people might be a little reluctant in front of the camera to speak out openly.

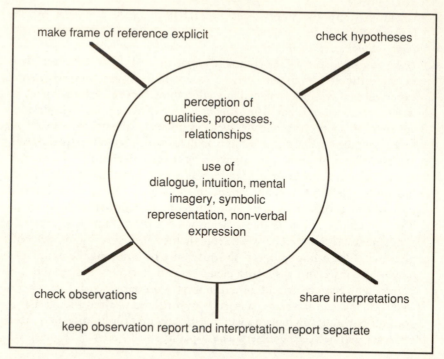

Figure 14.1 Methodology in biographic research

Note: The 'instrument', represented by the circle, is awareness of subjective experience. Around the circle, procedures to enhance validity constitute a 'safety frame' to avoid circular reasoning.

An audio-tape is less interfering. In order to make it scientific one should make the selection cues explicit.

The third criterion is transparency of interpretation. This is the question of how to avoid *circular reasoning*, i.e. to avoid the researcher drawing conclusions on the basis of pre-existent concepts he himself has imposed upon reality. A 'safety frame' has to be drawn. (see Figure 14.1).

A possible solution is to make the frame of reference one uses explicit. Often discussions in medical journals about supposed influence of 'mind' on somatic healing are controversial because there is (unknown) disagreement or confusion about the chosen frame of reference. No scientific observation or interpretation is possible without a frame of reference. One can choose whatever frame is preferred, whether it is biophysics, chemistry, biology or psychosynthesis, neuro-linguistic programming, or psychoanalysis, as long as it is explicitly stated. Only this can make comparison possible.

A second suggestion is to keep observation report and interpretation report separate. In order to render an observation report open to others, it has to be made independent of personal interpretations. An interview should be recorded and the text typed. In a column next to the text, non-verbal expressions can be registered. After this primary registration, the process of interpretation can proceed in two phases. First the experiences the interviewer had himself during the interview are written down. Afterwards all interpretations made after the interview can be added in a separate column.

Next the observation report can be shown to the interviewee to see whether it is congruent with his/her true experience.

It is also helpful to form a group of researchers to train each other to improve their perception of 'what really is', to make each other aware of the ever-present traps of bias and circular reasoning, and of the frames of reference, which are unconsciously used. This is called *triangulation* in social research.

To make the reasoning clear to readers, every step taken must be made explicit:

- Why was this question important?
- What did I experience myself when he said this?
- What do I still need to know?
- What hypothesis can I draw from what I know already?

If every statement is accompanied by an awareness of its origin, the whole process of observation becomes transparent to other observers. This does not mean that they have to agree with the kind of interpretations that are made, but at least a stranger will know where the statements come from. Computerized text analysis greatly facilitates this method of qualitative research and is a promising field.

9 IN CONCLUSION

Both scientific inquiry and the health-care system are searching for new ways to deal with the health problems modern industrialized societies are confronted with: for example ageing, cancer, chronic disease, and stress-related disorders. A biographic approach can throw new light on aetiology and context of disease. It is also a suitable method for investigating how personal processes affect health. Because biographic research looks at different levels of human life (medical, social, intrapersonal, interpersonal, existential), it could give an answer to the 'how and why' some people have a strong sense of coherence and, hence, good health. In this way biographic research can deliver very useful information for health promotion: e.g. do we have to invest so much energy in prevention campaigns against cancer which only focus on stopping smoking, changing diet and regular medical check-ups? Biographic research at least suggest it is equally important to pay attention to existential qualities. It is time that the subjective experience of people is no longer denied in scientific dialogue, but is taken seriously.

However, high-quality research is required in this scientific venture. An effort to improve systematic and careful observation is necessary along with a willingness to meet the other person as a human being. Researchers will have to be trained both in using the frames of reference which are suitable for biographic research and becoming good participating observers. Only then can a real and fascinating dialogue begin. Biographic research will have an undeniable place in the health sciences of the future.

ACKNOWLEDGEMENTS

We are grateful that this research project has been made possible by a donation of the ASLK–CGER–Bank of Belgium.

REFERENCES

Andrews, V. H. and Hall, H. H. (1990) 'The effects of relaxation/imagery training on recurrent aphtous stomatitis: a preliminary study', *Psychosomatic Medicine* 52: 526–35.

Antonovsky, A. (1987a) *Unravelling the Mystery of Health: How People Manage Stress and Stay Well*, San Francisco/London: Jossey-Bass Publishers.

—— (1987b) 'The salutogenic perspective: toward a new view of health and illness', *Advances, the Journal of Mind–Body Health* 4(1): 47–55.

Assagioli, R. A. (1965) *Psychosynthesis*, New York: Viking Press.

Cousins, N. (1976) 'Anatomy of an illness as perceived by the patient', *The New England Journal of Medicine* 295(26): 1458–63.

—— (1979) *Anatomy of an Illness as Perceived by the Patient*, New York: W. W. Norton.

Fiore, N. (1979) 'Fighting cancer – one patient's perspective', *The New England Journal of Medicine* 300(6): 284–9.

Glaser, B. and Strauss, A. (1967) *The Discovery of Grounded Theory: Strategies of Qualitative Research*, Chicago: Aldine.

Groddeck, G. (1930) 'Schweninger', *Der Artzt* 2: 167–74.

—— (1983) *Krankheit als Symbol, Schriften zur Psychosomatik*, Wiesbaden.

Grossarth-Maticek, R., Bastiaans, J. and Kanazir, D. T. (1985) 'Psychosocial factors as strong predictors of mortality from cancer, ischemic heart disease and stroke: the Yugoslav prospective study', *Journal of Psychosomatic Research* 29: 167–76.

Huygen, F. J. A. (1982) *Family Medicine, the Medical Life History of Families*, New York: Brunner/Mazel Publications.

Lafaille, R. and Lebeer, J. (1991) 'The relevance of life histories for understanding health and healing', *Advances, the Journal of Mind–Body Health* 7(4): 16–31.

Lerner, M. (1985) 'A report on complementary cancer therapies', *Advances, the Journal of Mind–Body Health* 2(1): 31–43.

Leroi, R. (1977) 'Resultate bei der Iscador-Nachbehandlung des operierten Mammakarzinoms aus der Lukas-Klinik', *Mitteilungen der Verein für Krebsforschung* 9(3): 16–26.

Locke, S. E., Kraus, L., Leserman, J., Hurst, M., Heisel, J. S. and Williams, R. M. (1984) 'Life change stress, psychiatric symptoms, and natural killer cell activity', *Psychosomatic Medicine* 46: 441–53.

Lusseyran, J. (1965) *And There was Light*, Edinburgh: Floris.

Lynch, J. J. (1982) 'Interpersonal aspects of blood pressure', *Journal of Nervous and Mental Diseases* 170: 143–53.

Miles, M. B., Huberman, M. A. (1984) *Qualitative Data Analysis*, London: Sage Publications.

Osler, W. (1921) *The Evolution of Modern Medicine*, Newhaven: Yale University Press.

Paget, Sir J. (1870) *Essays and Addresses by Sir James Paget*, London: Longmans, Green & Co.

Prigogine, I. and Stengers, I. (1984) *Order Out of Chaos*, Toronto: Bantam Books.

Ricoeur, P. (1955) *Histoire et Vérité*, Paris: Seuil.

Sabom, M. (1982) *Recollections of Death*, New York: Harper & Row.

Spence, D. P. (1982) *Narrative Truth and Historical Truth: Meaning and Interpretation in Psychoanalysis*, New York: W. W. Norton.

Spiegel, D., Kreamer, H. C., Bloom, J. R., Gottheil, E. (1989) 'Effect of psychosocial treatment on survival of patients with metastatic breast cancer', *The Lancet* 2: 888–91.

Tilkin-Franssens, D., in R. Lafaille, J. Lebeer and D. Tilkin-Franssens (1991) *Some Current Scientific Models for Biographical Research in the Health Sciences*, prepublication no. 4, Antwerp: International Network for a Science of Health.

Vansina, J. (1972) *Oral History: A Study in Historical Methodology*, London: Routledge & Kegan Paul.

Weizsäcker, V. von (1957) *Pathosofie*, Göttingen: Vandenhoeck & Ruprecht.

15

THE MEANING OF PERSONAL
EXPERIENCE FOR A SCIENCE
OF HEALTH

Janin Vansteenkiste

1 INTRODUCTION

From the time that rationalistic, medical science first emerged in Western civilization (with the School of Hippocrates, who rejected the older 'magical' ways of healing; see Lindeboom 1971; Papadakis 1978), a split has grown between personal experience and rational-scientific knowledge. Alongside this theoretical distinction grew, on the social level, a split between professional knowledge and the experiences of 'lay' people (Vansteenkiste 1978; Stacey 1988). In this contribution I will reflect on some of the consequences of ignoring, in the medical and health sciences, personal experience as a source of knowledge, wisdom and action. I will not deal with these questions here from an abstract, theoretical point of view, but try to draw some conclusions from actual life experiences.

This contribution is based partly upon my work in programmes for personal development, called biography programmes (BP). These programmes aim at those who are perhaps under high levels of stress, but whose general health allows them to take on an intensive self-development effort. It is also based on work in Namaste House[1], a temporary residence for persons who suffer from chronic illness or pain, from tensions, exhaustion, or diverse physical complaints. These guests find the opportunity to recover and to give attention to their life, aided by therapists, social workers, physicians, nurses, massage and physical therapists, etc. There are two types of residence. Guests can take part in a group programme of health development, structured around the physical ailment or illness. Alternatively, guests can develop an individual programme to encourage self-healing.

The value of biography programmes for research is that participants provide feed-back about the 'healing influence' they have experienced. Their feed-back is often spontaneously phrased in 'health promotion' terminology. Participant feedback is not solicited or systematically studied. However, the tendency of participants to offer health-related commentaries is consistently observed by professional practitioners in biography work (*Tijdschrift voor Sociale Hulpverlening vanuit de Antroposofie* 1987). Enquiry into what is being referred to by these healing influences, and the relationship between the participants' self-development and their notion of health, is valuable for our understanding of health. Further, it may be worthwhile to investigate the health-promoting potential of this type of personal development work.

In addition, severely ill, sometimes terminally ill, guests at Namaste House achieve a quality of life, a vitality and a self-determination, which they value highly. It would be valuable to investigate, from a science-of-health point of view, the nature of the healing processes in such cases.

2 SOME GENERAL BACKGROUND INFORMATION ON THE REPORTED EXPERIENCES IN BIOGRAPHY PROGRAMMES

The intellectual root of these programmes is the scientific contribution from Bernard Lievegoed (1978, 1983).[2] Lievegoed's insights on human biography are founded on his work as a psychiatrist. In his book *De Levensloop van de mens* (The Human Life-Cycle, 1978), he describes a subjective, individualistic viewpoint of the human biography. A human being is seen as a threefold being: body, mind and spirit. The human biography, as a unique personal 'work of art', can only be truly valued when these three elements are combined into one image. He stressed that it is always fruitful to situate a person's actual problems in the totality of the life-span: the present is as much determined by the past as by the future. Determinants of the past are fixed, but, as the present unfolds, it leaves freedom for action. It is always more important to find a new future than to linger too long in the past. Though there are transcultural implications, the appeal of individual biography work is mostly to people with a 'Western' cultural orientation. An explanation of the extreme individualization processes in Western cultures is described by Lievegoed in a more recent study, *De Mens op de drempel* (Man on the Threshold, Lievegoed 1983).

The biography programmes often address questions of vocation and orientation. The process is preferably conducted for small groups of participants (maximum 6 persons at a time). There is a process of mutual selection, prior to participation. A necessary principle in biography programmes is what I call 'freedom of development'. Personal development can be supported and encouraged, but it cannot be enforced. Participants know that the programme is intended to open up possibilities but that they will retain 'ownership' of their biography management. This meets the expressed need of participants, which is 'not one for an authority or expert who will tell them what to do' but for a work-form, a 'method' that will allow them to assess their life in a comprehensive, structured and meaningful way.

Invariably – in spite of doubts and anxieties – people who enrol are those who are not just discontented with their life: they have realized that change requires a personal effort and they are ready to take matters into their own hands.

3   VOCATIONAL QUESTIONS, CAREER PROBLEMS AND HEALTH

Participants initially express general complaints of dissatisfaction with their life. Under the cover of these introductory statements lie problems at work or inner conflicts about worklife. Typical descriptions, which recur over and over, sound like these:

- 'I cannot see myself doing the same things for the rest of my career.'
- 'I hate my job.'
- 'My work is boring, I wish I had an alternative.'
- 'I've come to a standstill: I'm not learning anything new.'
- 'I don't see the meaning of what we are doing here.'
- 'This work situation is just too demanding, I'm not sure if I can cope.'
- 'The stress at work has become unbearable.'
- 'Some of us will be made redundant soon, I wonder what else there is for me.'
- 'I never wanted to become a . . . (teacher, nurse, . . .).'
- 'I don't want to sacrifice my home life for the sake of my job anymore.'
- 'I just keep working for the money, the job is totally uncreative.'

It is well known that inadequate coping with stress, meaninglessness, uncertainty, stagnation, and lack of fulfilment, play a determining role in illness and disease. Many of the people I work with show minor, psychosomatic complaints. It will be clear that these express their current life situation (Van der Stel 1987). Reflecting on their biographical development and guidance by their biography programme, participants comment on the beneficial, empowering influence they experience. Many participants report that the biography programme has helped them to come to a new insight into long-standing problems: new light on 'old fact' lifts the burden of the past. For example one person suffered from the childhood memory of a cold upbringing without much parental presence. The discovery of how surrounded she was by loving persons outside the family took this discontentment away. Another person reported a major breakthrough when he came to understand why he had developed a particular handicap (blindness) in his life, what new human qualities this required from him and how these contributed to his actual profession.

Several participants reported that when they realized how their patterns of decision-making continually led to all kinds of obstacles and disappointments they could stop seeing themselves as victims of circumstance. A sense of a potential to influence the course of events set in afterwards. A similar

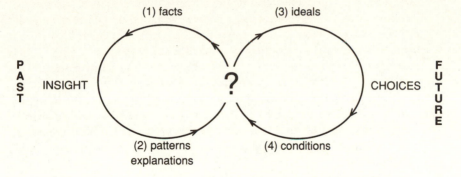

Figure 15.1 Research route

kind of relief was experienced by persons who observed that they seemed to have developed a certain illness, each time their life was at a crossroads.

For nearly all participants, putting intensive energy into recalling and describing their past has a 'metabolizing effect': old convictions and habits can be experienced as having served their purpose (or time) and space is created for new, more wholesome and innovative ones. Having to weigh up one's life achievements and (maybe undeveloped) talents, brings about a more balanced self-perception and more self-esteem (e.g. 'I thought I was never especially good at anything but I realize that I have to improve my communication skills', or 'I have never used my real talents! I am now more in touch with what it is that I want to do and need, rather than just see others' expectations'). Most participants report specifically the joy, warmth and courage it gave them, and a renewed sense of self-worth, a sense of direction. It is a strong motivating element towards starting new habits. It is not unusual, during or after a biography programme, to notice that a person starts taking better care of her/himself, crosses barriers of fear or develops patience. Persons let go of superfluous or stressful activities or introduce new ones. Quite often initial complaints of sleeplessness, anxieties, restlessness, breathing problems . . . disappear at the same time.

4 THE BIOGRAPHY PROGRAMME CONTENTS

Participants follow a research route, which takes them from their present situation, to their past, into their future, and back to the present (Figure 15.1). Each person sets off with her or his question of departure: the question is the steering wheel for the journey (Schöttelndreier 1989). Interestingly enough the starting question, which initially brought the person into the programme, undergoes some metamorphosis throughout this process.

- In Part (1) the individual past is retraced by recollection of memories and their past life. The attitude which is encouraged here, is one of 'looking back', of remembering, and leaving out judgements and rationalizations. Listeners who mirror the life story being told to them provide active but non-judgemental feed-back.

- Part (2) is a second life review, this time according to instructions to distinguish between what happened in private life, personal development and work-life. The life-pictures that emerge from this ordering of facts can point to particular connections, interrelations, and recurrent features in a person's life: it is part of researching which life-themes are held by someone, and – underneath these – what life-task or mission is implied.

- To build on the previous step, each person's ideals are explored in Part (3): they are seen as clues to the future. Both near and distant future are explored. The attitude with which this part of the research is performed, now shifts from thinking to relying on intuition and its antennae.

- In Part (4) the link is made between future choices and present conditions and constraints: which conditions are to be met in order to pursue a particular life course. This activity results in long-term aims and a short-term plan of action.

5 FEED-BACK FROM BIOGRAPHY PROGRAMME PARTICIPANTS

Participants usually evaluate the outcome of the process as positive: but their original problem is not necessarily solved. Often it is replaced by a 'new' question. In the course of the programme most participants discover that a more fundamental question is hidden under their initial problem. They then see that it is important to try and solve this deeper question before any changes can take place on the surface layer. They express surprise at 'how much there was behind the initial question'. Participants report that it has been a freeing experience to have been listened to, without being judged, and this unburdening was a relief. The power of the programme is in working with the future forces, according to participants. This is in contrast to the therapeutic focus on the past. 'I don't want to be analysing the past any more, I want to go forward', is a phrase often heard in this respect. Participants appreciate not being identified in a negative sense with their problem (not being seen as a problem-person but rather as a person who has a problem). They appreciate working from a concept of human development where crises and problems are an accepted part (no development without crisis).

6 RELEVANCE OF BIOGRAPHY PROGRAMMES FOR A SCIENCE OF HEALTH

Why do participants assess their experience of biography work as 'healing'? There is the obvious association with having experienced one's biography as 'a wholeness'. But there is another aspect: participants refer particularly to their experience of 'significant sharing'. Time and again, I observe in these programmes how much inner isolation people endure in daily life; what chilling, hardening and paralysing effects modern life has, in the social realm. And how deep the needs are for people to break out of that and to rediscover solidarity with others, a 'social bond', the warmth and creativity streaming from real meetings and from 'doing things together'. The need for healing social connections in our culture is paramount. A clear conclusion to emerge from exploring human biography is that health and healing are largely in the social realm. Seen in this light, a scientific basis for the contemporary health issues needs to be 'sociological' or 'cultural' in the broadest sense. The fundamental directions to be followed in healing transcend the scope of biomedicine.

Another important dimension is that participants are not under the pressure of 'an expert opinion' but are placed in a situation of equality. Participants often express that 'the last thing that they want, is another diagnosis'. This observation may help us to reassess the social status of scientific authority in the medical and therapeutic field. To what extent are the belief-systems in the achievements of science, the status of expert opinion and 'professional dominance' obstacles to attitudinal change? How is a science of health to be in tune with changing perceptions and emancipatory trends? Can it fulfil an enlightening and helping function in society? How?

Witnessing the spontaneity with which people take to the concept of 'biography', I see a need for a renewed concept of human life: a concept based on a picture of development in the human life-cycle. People do not experience the concept of biography as a technical model. They adopt the concept of biography because they experience it as a living force: it is apparently a concept which is effective in galvanizing them into motion. This points to a task, for a science of health, to conceptualize the human biography and to do it in such a way as to generate 'living ideas' rather than merely to construct theoretical models.

The biography programmes take place, not in a context of health promotion, but in a context of self-development, self-empowerment and self-management. Human life is the frame of reference, within which problems (including illnesses) can be given meaning, and thus integrated and acted upon. In this realm of work, people increasingly claim or adhere to their right of self-determination, the right to own their self-image, and (whether in health or sickness) the right to do the ultimate decision-making about their life-problems, including matters of health and death. People accept guidance in the construction of their self-image but they do not want determination from outside. This attitude allows an experience of 'healing' as a self-development terrain. Participants have chosen a 'label-free' setting and least of all do they want to be put in a 'patient role'! The implications are that health or medical research and practice must not 'problemize' or 'medicalize' human beings. The emancipatory self-image, the individual right to self-determination could be at the core of the healing arts and the science of health. 'L'art de guérir a toujours été le reflet de l'image que l'homme se faisait de lui-même . . .' (Bott 1980).

In the biography programmes it makes sense for the participants to view their particular health problems within the broad context of their life history: 'What does a particular health problem mean in my life?' This results in conclusions which are quite different from those reached through the usual logic of seeking those life factors which cause an illness. Participants gain insight into their personal obstacles in health when they are looked at from the totality of their developing biography. For this reason, the biographical perspective could be immensely helpful in understanding the aetiology of diseases that are mysterious to science, and could aid the treatment of diseases such as asthma, cardiovascular disease, cancer, etc., 'life-style diseases' that modern medicine cannot cure.

7 SOME GENERAL BACKGROUND INFORMATION ON NAMASTE HOUSE

The priority of Namaste House is to offer the guests, who arrive with their pain, illness and distress, a daily life, which invites them to dwell upon their thoughts, feelings, longings and hopes. The aim is to provide support as people connect their inner experiences with the circumstances and facts of their life (Van den Eynden 1992). People asking for residency are motivated by their need to stand still and stop so as to face their illness. They are on a quest for inner peace and rest, and expect to find in the house a context that upholds these qualities. The team of co-workers there are oriented towards a holistic, ecological view on health (Rijke 1989; de Vries 1985). They share the belief that within each person resides a capacity to heal; a capacity seldom fostered by mechanistic medicine and its treatment by medication.

8 HEALTH PROBLEMS IN THEIR LIFE-CONTEXT

Guests invariably do bring with them major illnesses, ranging from migraine to multiple sclerosis and cancer. They arrive, each at their own stage of inner assimilation of their condition. They may have gone through much on the physical level, but inside, in their thoughts, feelings and will, there is disruption. The inner healing processes still have to take place and still need to be supported. This level of healing starts with confronting the problem. One of the guests arrived, after being dismissed from hospital after mastectomy. 'At the hospital I was regarded as a model patient,' she said, 'but now at home it is as if the walls are coming down on me. I have to get out.' This woman's process of healing was characterized initially by a relatively long period during which she chose to relate to others on a friendly basis, just communicating about practical matters. Crucial periods followed when she expressed her aggression. In her own time, by stages, she came to the assimilation of her health problem. Assimilation, not in the sense of solving the problem: clearly some facts cannot be changed. Inner assimilation changes the way of seeing the problem and of relating to it.

A decisive breakthrough in assimilating takes place when a person is able to give meaning to his or her problem. This meaning emerges out of a personal development process: the person learns from the pain, for the future, to take responsibility for the effects of his/her own way of being. The person learns to influence his/her relational context. A woman suffering from multiple sclerosis, barely able to move and to communicate, came to the house, hoping to find meaning. She revolted against her dependency on help from others. This woman worked for a long time to acquire 'free inner choice-making' about what kind of care was to be administered to her and the timing of help. When she returned for another stay at the house, she explained that she wanted to continue to work on her character and to learn to enjoy simple things. She formulated painstakingly on her letterboard: 'I don't let myself be dominated by my MS.' This was her moment of breakthrough. Her energy to live was raised. She acquired 'meaning energy' and 'enjoyment-energy' and experienced a dynamic healing process, in spite of further physical deterioration.

The human organism contains knowledge of the correctness or incorrectness of its functioning. If a person can connect to this feed-back, his or her inner wisdom can become the compass indicating the direction toward healing. A clear discovery of this was made by a person who came to recover from cardiac-infarction. The person felt intuitively that a new life-phase was beginning, and also was in fear of a second infarction. During her first six-week stay, she endured a lot of 'heart-pain'. She had long suppressed a great deal of grief for past events, and stress upon her heart had built up until the heart gave in. This person could read the compass of her heart and began to work through her past. She realized for herself the inner peace, which can carry and assimilate pain. Taking the compass function of the organism seriously generates in people the power to take their life in their own hands. Through this power, people find – within their specific life-context – autonomy and self-responsibility. At a deep level healing has occurred.

9 FEED-BACK FROM NAMASTE HOUSE GUESTS

Guests report that the Namaste set-up provides the conditions that are required for healing processes to happen. A great help toward self-acceptance is to experience that one is totally accepted by others, as a person, and as a person with problems. In addition to self-acceptance, guests underline that they feel more able to 'allow things to happen': they have increased their capacity for functioning as an 'open system'. Although the problem is not solved, the inner compass begins to work and they find their way of dealing with the problem. Guests also report that something has changed in their outlook on 'health'. They say that they no longer see health primarily as absence of disease or physical ailment. Now they see it as an open way of responding to life, allowing feed-back and integrating this feed-back. In their own words, guests describe how, as a result, they have disidentified with their pain, illness or handicap. They mention an improved capacity to manage, within the context of their life, and vitality. They especially report that they get more in touch with the dynamic character of life and of healing. Healing processes are seen as quite distinct from treatment processes.

10 RELEVANCE OF EXPERIENCES AT NAMASTE FOR A SCIENCE OF HEALTH

The guests at Namaste are afflicted with severe, often incurable conditions. Yet they experience healing. It is significant that these people, each in their own way, somehow transcend their medical condition and allow for an inner reality. This inner reality, whether it is named 'spiritual' or not, becomes alive and resourceful for them. The existence of an emancipatory trend towards self-determination, clearly perceived in the description of the biography programme participants, is confirmed by the Namaste House guests. Also, here, the freedom from pressure of therapy and the equality of contact between guests and co-workers are crucial factors. Maybe the co-workers at Namaste show professional attitude and behaviour, which is in tune with this emancipatory trend, among persons who would otherwise be treated as patients. Their professionalism focuses on enlarging the guests' consciousness of their healing processes. In addition they make it their task to guard a context in which respectful communication is enhanced.

With regard to a scientific conceptualization of 'biography', the experience of the Namaste guests underlines that healing comes about from building a bridge between one's outer and inner biography. A scientific conceptualization will have to do justice to both dimensions: if limited to symptoms or outer biographical facts, it will not understand the real sources of health and healing.

The experience of healing as a self-development terrain, described in the case of biography programme participants, finds confirmation amidst

Namaste guests. The observation that even those who are actually ill and vulnerable mirror the self-determination claim of the healthy, makes us re-examine what we mean by 'well-being'.

In the case of the biography programme participants, the importance of a biographical perspective to the study of health and illness emerged. The case of the Namaste guests points out how fundamental it is, for scientific development, to start from acknowledging the existence of natural healing processes. These existential healing processes are closely linked to the individual life-context. Though they seem to follow certain rules, they develop in unpredictable patterns. In order to avoid reductionism, our modern scientific minds have to come to grips with existential phenomena of such a complex nature.

11   RECOMMENDATIONS FOR THE FUTURE OF A SCIENCE OF HEALTH

Certain conclusions can be drawn, relevant to the development of a science of health.

So much can be learned from a single life. This is of major importance to science, providing it does not fall into the trap of generalizing factors out of individual histories and overlooking the uniqueness of individuals. The necessity of reconciling generalities with the perception of uniqueness of each individual is not new in medical science and practice. However, it becomes acute when a biographical perspective is incorporated into a science of health. It sharpens the issues:

1   On the level of theory: how to generalize information obtained from individual biographies.
2   In healing practice: how to couple a general insight into the human biography with the 'healing intuition' of the unique therapeutic need of each person.

The first issue can perhaps be resolved on a methodological level. The second one challenges healers/practitioners to train and refine a suitable 'sense organ'. This sensitivity, in its turn, may well be trained by an ongoing study of individual biographies and case studies of individual outcomes of treatments (why was that person helped, or not helped, with a treatment of that kind?).

A major research question is to investigate *what mobilizes the human will*. Science is still in the dark about what energizes and mobilizes the forces of the will. Yet, the question of what brings a person into motion is the basis for our healthy life-styles, healthy cultures, for coping with illnesses and for appropriate ways to die. Working with the riddle of the will would lead research beyond the perspective of medical science, towards a deep understanding of the human being.

Academic institutions are still too far away from 'real' life. Existing academic standards still reject or at least subordinate personal experience as a valuable source of knowledge. The old split between personal and scientific knowledge remains. Although many academics recognize that there is a problem in this field, initiatives towards a solution of this cultural split are hard to find. This is particularly relevant for the health field for *many health practices have their foundation in personal experience* (relaxation, body awareness, meditation, etc.), which ought to be complemented by a more systematic knowledge. Many interesting movements are developing outside traditional academic institutions, such as new psychotherapeutic techniques, collective awareness in self-help groups, the development of self-care techniques, new methods of leadership, and so on. Developments of this kind are of prime importance for health sciences in the future. One can envision a strategy in which a permanent dialogue between all relevant sources of knowledge becomes institutionalized as an indivisible part of the development of a science of health. Without dialogue, without honouring personal experience, I cannot see a science of health which truly holds the transforming power to heal.

REFERENCES

Bott, V. (1980) *Médicine Anthroposophique*, Paris: Triades.
Lievegoed, B. (1978) *De Levensloop van de mens*, Rotterdam: Lemniscaat.
—— (1979) 'The phases of human life', in K. E. Schaefer, U. Stave and W. Blankenburg, *Individuation Process and Biographical Aspects of Disease*, New York: Futura, pp. 109–26.
—— (1983) *De Mens op de drempel. Mogelijkheden en problemen bij de innerlijke ontwikkeling*, Zeist: Vrij Geestesleven.
Lindeboom, G. A. (1971) *Inleiding tot de geschiedenis der geneeskunde*, Haarlem: De Erven F. Bohn.
Papadakis, Th. (1978) *Epidauros: The Sanctuary of Asclepios*, Athens: Meletziz & Papadakis.
Rijke, R. and J. (1989) *Zingeving bij ziekte en gezondheid: van een mechanistische naar een ecologische visie*, Deventer: Van Loghum Slaterus.
Schöttelndreier, J. (1989) *Levenslijnen. Levensvragen, crisismomenten, toekomstkansen*, Zeist: Vrij Geestesleven.
Stacey, M. (1988) *The Sociology of Health and Healing*, London: Unwin Hyman Ltd.
Tijdschrift voor Sociale Hulpverlening vanuit de Antroposofie (1987).
Van den Eynden, D. (1992) 'Namaste-Huis. Het gaat om het ondersteunen van helingsprocessen' (The Namaste House. Supporting Healing Processes), *Sociaal*, 5 May.
Van der Stel, A. (1987) *Eigentijdse ziektebeelden*, Zeist: Vrij Geestesleven.
Vansteenkiste, J. (1987) *Institutionalisering van de medische professie*, Brussel: Nationaal Onderzoeksprogramma in de sociale wetenschappen: Eerstelijnszorg, 1A.
Vries, M. de (1985) *Het behoud van leven*, Deventer: Bohn, Scheltema & Holkema.

Part V

CONCLUSIONS

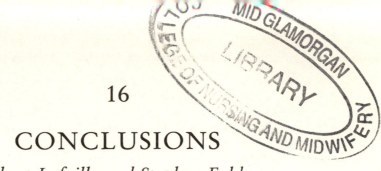

16

CONCLUSIONS

Robert Lafaille and Stephen Fulder

Most of us would agree that the burden of illness within our modern world is both excessive and unnecessary. The major health problems, such as cancer or heart disease, are rooted in our actions, our culture, our environment and our community, and we have so far failed significantly to reduce this burden. It is becoming clear that only a shift in emphasis towards promoting health rather than curing disease will have a significant impact on these problems. This shift should be part of both national policy and individual attitude and behaviour.

However, such a historic shift in our cultural perspective requires a deeper understanding of what health is. So far such an evolving health perspective has grown as a subculture phenomenon through self-help groups, complementary medicine, holistic clinics and personal health consciousness. It has made little headway in the academic world. For this reason the 'Think Tank' on the New Science of Health was held at Louvain-la-Neuve, and this book was compiled.

Historical and cultural enquiry, some of which is presented in this book, has shown us that the way we view health is a bit like the way we view goodness or values. Namely it is completely tied in to the underlying *Weltanschauung*, the assumptions, archetypes and cultural premises of the times. The view of health that we have had for the last two centuries is inextricably connected to a mechanistic, positivistic and reductionist metaphysical position and worldview. As contemporary science became a dominant way of understanding the universe, so orthodox medical attitudes to health dominated. Health and disease became a matter of discrete physical signs and symptoms that were fixed in time and space. Statements about health were only valid if they were provable by conventional scientific methodology. Some of the features of this thinking on health have been described in this book, and we can summarize them here. Health is:

- *Dualist*, because on the one hand the body is studied as a mechanically working device, on the other hand the mental phenomena are seen as something of fundamentally different nature.

237

- *Discrete*, because it is assumed that there is either health, or the absence of it. Even in the expanded WHO definition, this *all-or-none* criterion is often used in that sense: health is equal to physical well-being, general well-being, psychological well-being or social functioning; disease is the absence of these.

- *Abstract*, in the sense that the study of disease in many disciplines is often far removed from the everyday reality and the experiences of living people. Often in the study of pathology, or in the study of the risk factors and health, the subjective experiences of people are left out, as if they did not matter.

- *Static*, in the sense that health is assumed to have had the same meaning in history, to have the same meaning within different people, or in the course of a single life history, and it is assumed that it will ever remain so.

- *Deterministic*, in the sense that the universe, the human body, or human relationships are seen to be essentially governed by eternal laws which can lead to predictable outcomes.

- *Linear-causal*, in the sense that health is determined by a cause/effect relationship: it is assumed that diseases must have detectable causes, which can and have to be eliminated. And it is assumed that health promotion is a collective or an individual effort to fight the causes of disease, or to increase material resources, physical fitness and perfection, social security or social support.

Times are changing and we are entering a new era where the cultural reliance on the safety of a clockwork universe is coming to pieces. In all the fundamental sciences, from quantum physics to astrophysics, and in the humanities from psychology to sociology, the mechanistic viewpoint is now seen as inadequate and questionable. In this book we have seen that there are other ways of looking at ourselves, through our health, which are potentially more fruitful. For example the integrative, multidimensional view, supported by systems theory. Or the ecological view, stressing relations, and processes rather than objects; or the subjective view stressing the experiential, personal dimension; or the social dynamic view emphasizing the social and political forces involved in shaping health. These new views, discussed in this volume, give us a much expanded picture of health.

Instead of dualism, it would be more adequate to speak of *integration or connectedness of physical and mental phenomena*. The distinction between somatic and psychologically based diseases is artificial. Health is constantly influenced by somatic and by mental processes, which are interwoven. To see health on a *continuum* means that health is a relative state: you are more or less healthy. On one side of the spectrum is 100 per cent disease, and on the other side is 100 per cent health. One can be healthy even in the presence of defects of the body. One can be unhealthy, even in the absence of bodily defects. Health is about life. The study of health must take place in a constant referral to the way of life of living people and their everyday *experiences*, in their environment. A *dynamic* state means that the experience and understanding of health is different in various epochs of cultural history. This does not mean that at any time there was a wrong view. It was simply different. In the same person 'health' can also have different meanings and criteria during the course of a lifetime. Disease is not the opposite of health, but may be seen as an integral part of health. The *indefiniteness* of health relates to recent insights in fundamental sciences such as quantum physics and the theory of chaos and order in complex unstable systems. It is a natural property of such systems to attain a higher level of organization in the midst of chaos, of which the outcome might be unpredictable. As life and the healing systems of all living organisms are also complex, unstable systems, so is the evolution of the body in case of disease and of health. This process is essentially undetermined and leaves room for freedom and creativity.

In this volume we have not attempted to produce a catalogue of definitions of health, old and revised, nor have we attempted to cover all the possible areas of redefinition. However, the various contributors, by giving their own 'exploded' view of health from within their own field, give suggestions as to how new meta-models can arise. This then raises the question, how is health to be studied under such academic frameworks? Obviously a new science of health will have to be inclusive, integrative and diverse. It ought to incorporate the study of:

- The health potential of people with a disease.
- The subjective experiences of people in relation to the objective properties of somatic processes.
- Qualities as well as quantities: e.g. human qualities like awareness, consciousness, autonomy, freedom, love, will, or qualities in biological systems like healing, autonomy, development.
- The time-dependent processes leading towards health, in individuals as well as in societies and in history.
- The interrelatedness of individuals, societies, or ecosystems in relation to health, e.g. social interaction, relationships between people and the relationship between health and power (political systems, health-care systems).
- The creative processes in individuals and groups in relation to health.
- The flexibility, the ability to change, learn and regenerate, of living organisms (cells, organ systems, individuals, and groups) in case of damage.
- The extent to which individuals and societies are vulnerable to, e.g. toxic or pathogenic factors or other risks.

Most of the contributors have stated the view that a future science of health will be a science of qualities. Not that it will be a normative science, but that researchers will relate their work to human values and social targets. Inevitably, it will also be a science which studies more complex and changing realities: the permanent changes and transformations that take place (historical, political, social, personal, biological, and so on). The notion that there is a simple causal relationship between certain factors and health, or that there would exist some simple rules to guarantee health, are notions which are contradictory to the current knowledge of the complex and dynamic processes involved in health and illnesses. Progress in methodology, epistemology and computer technology will support and enhance our ability to study complexity.

These conclusions are general. How do they translate to actual research programmes within an academic setting? To some extent new research ideas will emerge naturally as a result of the creative exercise of working within new paradigms. The clarification of concepts, the construction of meta-models or theoretical integration, biographical research, the study of complexity using field concepts or systems theory are all pathways to the study of health which should enrich or revitalize the field. Hopefully innovative research in the areas of health policy, epidemiology, biopsychosocial medicine and ecology, will emerge as a result. Some actual suggestions for research targets that would become the themes of a new science of health follow:

- Theoretical and historical research in the health sciences. An international research programme in this field could be very helpful.
- Research to assess and support human empowerment (participation, self-help groups, evaluation of the effectiveness of self-care techniques, etc.) and new social relationships within the health-care system and in society.
- Research into placebo effects, directed towards healing interaction, healing substances and the development of a more humane health-care system. Placebo-effect research could be helpful to reorder financial resources in the health-care system.
- Qualitative and biographical research which could deepen our understanding of human life experience. Themes like autonomy, awareness, meaning in life, spirituality, etc. are not epiphenomena for the development of the health sciences but have to be rescheduled as research topics of utmost importance.
- Research into the (re)distribution of health, wealth and welfare. This kind of research has to be linked to power processes. On a global scale, poverty and war are still the most widespread and influential risk factors for health. These threats to humanity can only be reversed by the redistribution of power.
- Research directed towards the installation of a new ecological equilib-

rium. This goes way beyond risk calculation, into policy issues and the possibilities inherent in utopian ideas.
- Multidisciplinary, problem-oriented research into biological processes, re-investigating human susceptibilities in a holistic manner.

A lot of work lies ahead. New research programmes will have to translate these initial ideas into more workable forms and orient the research towards practical problems.

The WHO Charter of Ottawa defines health promotion as the process which enables people to increase control over, and improve, their health to reach a state of complete well-being. It requests all health-care professionals to contribute to the pursuit of health and to reorient the health services towards health promotion. The Ottawa Charter calls on authorities and individuals to advocate the promotion of health and to set up strategies and programmes for health promotion, especially at an international level. Applied to researchers and professionals in the health sciences, the more general ideas of the Ottawa Charter are an invitation to scientists to commit ourselves to contribute to real health on a global scale.

NOTES

INTRODUCTION

1 The notion of 'science' is different in different countries. In some countries 'science' refers only to very basic knowledge (to fundamental sciences such as mathematics, biology or epistemology); in other countries 'science' refers to all sort of academic disciplines and their applications. In this tradition there exists a science of leisure time, a science of social security, a science of social work, etc. We have to be aware of these national and cultural differences.

Here the notion of science of health is used as a concrete utopia, a motivating idea to reorient the existing health sciences. Openness to other frames of reference and approaches is fundamental to this purpose. The aim is to develop more integrated models and empirical research within the different sciences which deal with health.

1 TOWARDS THE FOUNDATION OF A NEW SCIENCE OF HEALTH: POSSIBILITIES, CHALLENGES AND PITFALLS

* A more extensive form of this contribution is available as Pre-publication No. 6 of the International Network for a Science of Health.

I want to acknowledge all the participants of the WHO symposium in Louvain-la-Neuve for their comment on an earlier draft of this article. I also want to thank Prof. Dr H. Coenen, Dr L. Debaene, Dr J. van Dixhoorn, Prof. Dr J. Dumon, Drs H. Hiemstra, Prof. Dr M. Ingrosso, Prof. Dr H. Janssens, Dr J. Lebeer, Prof. Dr I. Maso, Dr P. Mielants, Prof. Dr J. Heyrman, Prof. Dr M. de Vries, Dr J. Vansteenkiste, for their comments on earlier drafts or parts of it and Drs G. Baretta, Mr J. Batenburg, Dr S. Fulder and Drs A. Tingloo for their assistance with the English language.

1 Since the epoch-making work of Thomas Kuhn the term 'paradigm' has become very common in all scientific fields. This has led to a multiplicity of definitions and meanings, that we cannot deal with in this context. Nevertheless, it urges us to specify the significance we will give to these terms and concepts in this contribution.

To develop a sound classification of scientific tenets, it is important to use a terminology which differentiates concepts as to their range and size. For our purposes it is not important at this moment if, and to what extent these pro-positions are empirically verified. A nominalistic criterion is sufficient: a 'scientific' proposition is assumed to be scientific (and thus belongs to our universe of propositions) when it is recognized as such by the relevant actors in society. We will use the following terms:

Worldview encompasses *scientific paradigms* encompasses *scientific theories* encompasses *scientific hypotheses*.

Thus a paradigm is part of a worldview and refers to a well circumscribed set of scientific theories. A theory is a definite and structured set of scientific propositions.

2 An exploration of the difficulties which may appear in the discussion about worldviews and scientific paradigms, can be illustrated by three exercises, which are frequently used in contemporary psychotherapy.

- In the first exercise, the therapist splits a group in two parts: one at his left and one at his right side. In the middle he puts a plate with the 'same' photo on both sides. But the photos are similar, some details are changed. The two groups look at the same photo, without knowing the changes between the two versions. The groups are asked, by the therapist, to describe what they see on the photo.
  In no time a fight emerges about the picture on the photo. As long as they are under influence of the suggestions of the therapist that both pictures are identical, the two groups believe that they are talking about the same reality (= the photo), there is a struggle about the definition/interpretation of this reality.
- In the second exercise people are asked to form couples. Person A is asked to defend his worldview (his belief or disbelief in God, his meaning of life, etc.). The experience is that people become very aware of their belief system and how they are emotionally bound to it. The personal belief system is defended against invaders or attacked with lots of energy. More than anything else people are fighting for their worldview and their value system. There exists strong resistance against every change in this belief system. A large amount of energy supports this system and when this system is changed, this energy has to be transformed.
- In the third exercise people are asked to observe other people and describe their observations. After this first description they are asked to evaluate what in their description is 'objective' and what is interpretation. From experience we know that the list of interpretations is always growing; there are little or no observations which are only 'objective'. The description of reality is first of all an interpretative process.

Contemporary psychotherapy offers a lot of such exercises by which we are able to highlight our hidden convictions and show how they work out in the social field.

3 An alternative terminology could be: microscopic/macroscopic, small/broad spectra, or deductionistic/holistic. When this last distinction is used, it has to be liberated from connotations (a hidden good/bad distinction) as found in the literature.

4 We use the term 'culturological' to point to a whole class of scientific disciplines or subdisciplines which all use the perspective of culture as a main frame of reference.

5 See, e.g. the following quotation of Hesse: 'Geister wie Abälard, wie Leibniz, wie Hegel haben den Traum ohne Zweifel gekannt, das geistige Universum in konzentrische Systeme einzufangen und die lebendige Schönheit des Geistigen und der Kunst mit der magischen Formulierkraft der exakten Disziplinen zu vereinigen' (Hesse 1972: 131).

2 HUMAN HEALTH AND PHILOSOPHIES OF LIFE

1 This chapter has been translated by Adelheid Baker.

4 RELATIONS BETWEEN A SCIENCE OF HEALTH AND PHYSICS FROM MIRCOPHYSICS TO COSMOLOGY

* I want to acknowledge Dr R. Lafaille and Dr S. Fulder for their comments on earlier drafts and their aid in collecting the literature and other material.

1 To be taken in the sense given in the chapter by Lafaille in this volume.

2 This means going from photons, electrons, protons, neutrons, etc. (studied by physics), atoms and molecules and their reaction (studied by chemistry), giant molecules (studied by biochemistry), cells (studied by cell studies), living creatures (studied by biology), individuals, small communities, states, confederal states, world state (studied by the human sciences like psychology, to politics and health care). One may add intermediate steps, subdivisions, etc. There are overlaps too (e.g. atoms belong to physics and chemistry; unicellular objects may be studied by biochemistry and by biology as well as by cellular studies). Moreover, there are links between various levels. For example, a poison is a chemical material but it clearly affects the health of humans (or other living creatures), acting at several levels in chain. Similarly a stone hitting a person may be considered as a physical process: one may describe the motion of the projectile and the collision by physical laws. However, it will usually affect the health or well-being of the victim. These examples of interrelation are fairly simple, but of course more involved cases may be considered, such as in medical physics.

3 Examples of such premises could be:
  i   Life should be meaningful.
  ii   One should be able to trust the community.
  iii The community should care and be cared for (social programme).
  iv   Diseases, hunger and suffering should be combated worldwide.
  v   Healthy nutrition and life-style, and prevention of any kind of pollution is essential.

The further development and axiomatization of principles of this kind in the health sciences implies:
  i   Search for new principles.
  ii   Formulating the principles concisely and adequately; possibly subsuming some of them as consequences (theorems) of others.
  iii Checking consistency between the principles.
  iv   Deducing a large number of theorems (consequences) from the principles and turning them into practice, e.g. by introducing them as political or social actions.

4 Many developments in mathematics can support the study of complexity in a non-classical way in the health sciences: topography, graph theory, non-linear and discrete functions, new solutions of differential equations, etc. Note that several fields are known in the literature under different names. For example, chaos theory and its mathematical foundations are referred to as: dynamical systems theory, catastrophe theory, chaos theory, etc. These different names can stress differences which are only interesting for experts. In this chapter, we shall use 'chaos theory' as a common denominator of all these developments.

5 Some illustrations will make this clearer. Suppose a person goes for a walk and a roofing-tile falls on his head. Of course according to Newtonian theory this is

perfectly determined since the beginning of time. But who is going to calculate this, who is going to predict this? The person might have returned on his steps just because he remembered something; he could have slowed down his pace to look at the window of a shop, etc. Moreover the tile stayed quietly on the roof for years . . . It seems hard to believe that anyone, even God himself, has a computer big enough to keep track of all this (Einstein said 'Nature integrates empirically').

The following illustration will confirm the previous one. Consider a spoon containing, say 9 cm^3 of water, which corresponds to half the number of Avogadro. Hence the spoon contains $3*10^{23}$ molecules of water. Suppose someone could count a molecule every second of his life, day and night; after one year he has counted about $3*10^7$ molecules. If he continues for one century he could count about 3 billion molecules. Let one billion Chinese join him, each one counting a molecule, every second, day and night then $3*10^{18}$ molecules are counted after one century. To count the full spoon one needs to repeat a hundred thousand times the 100-year experiment with the billion Chinese! Now each molecule has a place and a momentum. Even in Newtonian mechanics it is beyond the reach of any man or any computer to determine perfectly the evolution of any molecules over a period of only 1 second!

6   HEALTH AT SOCIAL CROSSROADS: THEORIES OF HUMAN INTERACTION AND A SCIENCE OF HEALTH

1 It is useful to define the following notions:

interaction = the continuous stream of (non)verbal information exchange
relation = the meaning people give to the interaction (business relation, sexual relation, etc.)
relationship = the shared meaning of being exclusively related to each other (kinship, friendship, etc.)

9   A THEORETICAL MODEL FOR HEALING PROCESSES: REDISCOVERING THE DYNAMIC NATURE OF HEALTH AND DISEASE

1 Parts of this text are reprinted from, by permission of the publisher, *Humane Medicine*, 1985.
2 Under 'Medicine' I include here, while recognizing their own identity, all fields in the health professions, such as nursing, physiotherapy, etc.

12   INTEGRATIVE, TRANSDISCIPLINARY RESEARCH METHODOLOGY: PRINCIPLES AND APPLICATION STRATEGIES

1 A paradigm is defined here as a system of thought, a way of ordering and simplifying the world's overwhelming complexity by making certain fundamental assumptions about the way the world 'works'. Paradigms are normative; they determine what one views as important and unimportant, reasonable and unreasonable, legitimate and illegitimate, possible and impossible, and what to attend to and what to ignore. According to Thomas Kuhn (1970: 150): '. . . the proponents of competing paradigms ply their trade in different worlds. One contains con-

strained bodies that fall slowly, the other pendulums that repeat their motions again and again. One is embedded in a flat, the other in a curved, matrix of space. Practising in different worlds, the two groups of scientists see different things when they look from the same point in the same direction . . . and they see them in different relations one to the other.'

For these reasons, all explanations of problems and attempts to solve them are, ultimately, paradigm-based. Through the assumptions they embody, paradigms provide the basis for all normal scientific (problem-solving) activity, including areas of investigation, how the problem will be defined, what are considered to be data (and, by implication, what are not considered to be data), how data will be treated and interpreted, what conclusions will be drawn from the data, and what policy recommendations will be made.

14   REVEALING THE SELF: BIOGRAPHICAL RESEARCH AS A METHOD OF ENQUIRY INTO HEALTH

1 We only give a brief account of these life histories. The full data are available to researchers on request.

15 THE MEANING OF PERSONAL EXPERIENCE FOR A SCIENCE OF HEALTH

1 It is with the kind permission of Denise Van den Eynden, who shared her professional observations with me, that information on Namaste could be included for the purpose of this article. The Namaste House can be contacted at the following address: Namaste VZW, Centrum voor ervaringsgerichte gezondheids-ontwikkeling, Veldestraat 57, B-9850 Merendree (Nevele), Belgium.
2 The NPI Institute for Organization Development, founded by Lievegoed in Zeist (Holland) in 1954. Over the years a stream of organization development work has originated from this institute and spread worldwide, relying on practical concepts of the development of people and of structures. During the last 10 to 13 years, an explicit focus on biography work emerged. There is a strong creative link too between the biography work at the NPI Institute and the Centre for Social Development in England (Sharpthorne, Sussex). At this international training centre, biography work is furthered in the fields of adult education and counselling. Several other training institutes and consultancies form a growing network wherein working with the human biography (as well as the biographies of groups and organizations!) is practised. The common root to all this work is the inspiration of anthroposophy.

BIBLIOGRAPHY

1 HISTORY OF HEALTH AND THE HEALTH SCIENCES

Ackerknecht, E. (1973) *Therapeutics: From Primitives to Zen*, USA: Hafner Press.
—— (1982) *A Short History of Medicine*, Baltimore, MD: Johns Hopkins University Press.
Barkas, J. (1975) *The Vegetable Passion. A History of the Vegetarian State of Mind*, New York: Charles Scribner's Sons.
Benesch, H. (1990) *Warum Weltanschauung. Eine Psychologische Bestandsaufnahme*, Frankfurt am Main: Fisher Taschenbuch.
Garrison, F. H. (1929) *An Introduction to the History of Medicine, with Medical Chronology, Suggestions for Study and Bibliographic Data*, Philadelphia and London: W. B. Saunders.
Göpel, E. (1987a) *Lebensmodelle und ihre methodischen Konsequenzen für die Gesundheitsbildung*, Materialien des Oberstufen-Kollegs, Bielefeld: Oberstufen-Kollegs, Universität Bielefeld.
—— (1987b) *Beiträge zur Diskussion*, Materialien des Oberstufen-Kollegs, Bielefeld: Oberstufen-Kollegs, Universität Bielefeld.
Gurjewitsch, A. J. (1986) *Das Weltbild des mittelalterliche Menschen*, München: C.H. Beck.
Herzlich, C. and Pierret, J. (date n.a.) *Illness and Self in Society*, Baltimore, MD: Johns Hopkins University Press.
Imhof, A. E. (ed.) (1983) *Der Mensch und sein Körper. Von der Antike bis heute*, München: Beck-Verlag.
King, L. S. (1971) *A History of Medicine*, Harmondsworth: Penguin Press.
Lichtenthaeler, Ch. (1982) *Geschichte der Medizin*, 2 vols, Köln-Lövenich: Deutsche Ärtze-Verlag.
Meyer-Steineg, Th. and Sudhoff, K. (1965) *Illustrierte Geschichte der Medizin*, Stuttgart: Fisher.
Nitschke, A. (1989) *Körper in Bewegung. Gesten, Tänze und Räume im Wandel der Geschichte*, Stuttgart: Kreuz-Verlag.
Papadakis, T. (1978) *Epidaurus: The Sanctuary of Asclepios*, Athens: Meletziz & Papadakis.
Peeters, H., Gielis, M. and Caspers, C. (1988) *Historical Behavioural Sciences: A Guide to the Literature*, Tilburg: Tilburg University Press.
Peters, H. (1969) *Der Artzt und die Heilkunst im alten Zeiten*, Düsseldorf/Köln: Eugen Diederichs Verlag.
Pouchelle, M. C. (date n.a.) *Body and Surgery in the Middle Ages*, Cambridge: Polity Press.

247

Rosen, G. (1959) *A History of Public Health*, New York: MD Publications.
—— (1974) *From Medical Police to Social Medicine: Essays on the History of Health Care*, New York: Science History Publications.
Rothschuh, K. E. (1965) *Prinzipien der Medizin*, München: Urban und Schwarzenberg.
—— (1973) *History of Physiology*, UK: Krieger Press.
—— (1978) *Konzepte der Medizin in Vergangenheit und Gegenwart*, Stuttgart: Hyppokrates.
Schipperges, H. (1984) *Die Vernunft des Leibes. Gesundheit und Krankheit im Wandel*, Graz/Wien/Köln: Styria.
—— (1985) *Homo Patiens. Zur Geschichte des kranken Menschen*, München/Zürich: Piper.
Schmidbauer, W. (1973) *Psychotherapie: ihr Weg von der Magie zur Wissenschaft*, München: Nymphenburger Verlagshandlung.
Schumacher, J. (1963) *Antike Medizin. Die naturphilosophischen Grundlagen der Medizin in der Griechischen Antiken*, Berlin: Walter De Gruyter & Co.
Sigerist, H. E. (1956) *Landmarks in the History of Hygiene*, London/New York/Toronto: Oxford University Press.
—— (1961) *A History of Medicine*, vol. 2, *Early Greek, Hindu and Persian Medicine*, New York: Oxford University Press.
Starobinski, J. (1964) *A History of Medicine*, New York: Hawthorn Books.
Tannahil, R. (1973) *Food in History*, London: Eyre Methuen.
—— (1980) *Sex in History*, London: Hamish Hamilton.
Venzmer, G. (1968) *Fünftausend Jahre Medizin*, Bremen: Carl Schünemann.
Verbrugh, H. S. (1978) *Paradigma's en begripsontwikkeling in de ziekteleer* (Paradigms and Conceptual Development in General Pathology; in Dutch), Haarlem: De Toorts.
Wichmann, J. (1990) *Die Renaissance der Esoterik. Eine kritische Orientierung*, Stuttgart: Kreuz-Verlag.

2 MODELS AND THEORIES OF HEALTH

Foundations of a Science of Health (1989), report WH0–International symposium 'Toward foundations of a science of health', Antwerp/Louvain-la-Neuve, 19–24 March 1989 (R. Lafaille and J. Lebeer, rapporteurs).
Harmon, W. W. (1992) *A Re-examination of the Metaphysical Foundations of Modern Science*, Sausalito: Institute of Noetic Sciences.
Milz, H. (1985) *Ganzheitliche Medizin. Neue Wege zur Gesundheit*, Königstein: Athenäum.

2.1 Definition of health

Anderson, R. (1984) 'Health promotion: an overview', *European Monographs in Health Education Research* 6: 1–76.
Antonovski, A. (1987) 'The salutogenic perspective: toward a new view of health and illness', *Advances. The Journal of Mind–Body Health* 4: 47–55.
Capra, F. (1986) 'Wholeness and health', *Holistic Medicine*, vol. 1, pp. 145–59.
Engel, G. L. (1960) 'A unified concept in health and disease', *Perspectives in Biology and Medicine*, vol. 3, pp. 459–85.
Parsons, T. (1967) 'Definition von Gesundheit und Krankheit im Lichte der Wertbegriffe und der sozialen Struktur Amerikas', in A. Mischerlich *et al.* (eds), *Der Kranke in der modernen Gesellschaft*, Köln/Berlin: Kiepenheuer & Wich.

WHO (1988) *Basic Documents of the World Health Organization*, 37th edn, Geneva: WHO.

2.2 Paradigms/worldviews

Dijksterhuis, E. J. (1986) *The Mechanisation of the World Picture. Pythagoras to Newton*, Princeton, NJ: Princeton University Press.

Hampden-Turner, Ch. (1981) *Maps of the Mind. Charts and Concepts of the Mind and its Labyrinths*, New York: Macmillan.

Harman, W. W. (1992) *A Re-examination of the Metaphysical Foundations of Modern Science*, Sausalito, CA: Institute of Noetic Sciences.

Hedrich, R. (1990) *Komplexe und fundamentale Strukturen – Grenzen des Reduktionismus*, Mannheim.

Kuhn, T. S. (1973) *The Structure of Scientific Revolutions,* Chicago: University of Chicago Press.

Wilber, K. (1981) *No Boundary*, Boston & London: Shambhala.

Wildiers, M. (1984) *The Theologian and his Universe: Theology and Cosmology from the Middle Ages to the Present*, San Francisco, CA: Harper & Row.

2.3 Paradigms in the natural sciences and biology

Bohm, D. (1980) *Wholeness and the Implicate Order*, London: Routledge & Kegan Paul.

Capra, F. (1975) *The Tao of Physics*, Boston & London: Shambhala.

—— (1982) *The Turning Point: Science, Society, and the Rising Culture*, New York: Bantam Books.

Eccles, J. C. (1980) *The Human Psyche. The Gifford Lectures, University of Edinburgh*, New York/Berlin: Springer International.

Gleick, J. (1987) *Chaos. Making a New Science*, London: Penguin.

Goodwin, B. C. and Saunders, P. (1989) *Theoretical Biology: Epigenetic and Evolutionary Order from Complex Systems*, Edinburgh: Edinburgh University Press.

Goodwin, B. C., Sibatani, A. and Webster, G. C. (eds) (1989) *Dynamic Structures in Biology*, Edinburgh: Edinburgh University Press.

Ho, M.-W. and Fox, S. W. (eds) (1988) *Evolutionary Processes and Metaphors*, New York: Wiley.

Hofstadter, D. R. (1985) *Metamagical Themes: Questioning for the Essence of Mind and Patterns*, New York: Basic Books.

Prigogine, I. and Stengers, I. (1983) *Order out of Chaos*, London: Wiley.

Sheldrake, R. (1981) *A New Science of Life. The Hypothesis of Formative Causation*, Los Angeles: Tarcher.

2.4 Biomedical paradigm

Abholz, H.-H., Borgers, D., Karmaus, W. and Korporal, J. (eds) (1982) *Risikofaktorenmedizin. Konzept und Kontroverse*, Berlin: De Gruyter.

Badura, B. and Kickbusch, I. (eds) (1992) *The New Social Epidemiology*, Copenhagen: WHO.

Bakal, D. A. (1979) *Psychology and Medicine. Psychobiological Dimensions of Health and Illness*, London: Routledge.

McKeown, Th. (1979) *The Role of Medicine. Dream, Mirage or Nemesis?*, Oxford: Basil Blackwell.

2.5 Systems paradigm

Bateson, G. (ed.) (1975) *Steps to an Ecology of Mind*, New York: Ballantine Books.
—— (1979) *Mind and Nature*, London: Fontana Paperbacks.
Bertalanffy, L. von (1972) 'The history and status of general systems theory', in G. J. Klir (ed.), *Trends in General Systems Theory*, New York.
Callebaut, W. and Pinxten, R. (1987) *Evolutionary Epistemology. A Multiparadigm Program*, Dordrecht: D. Reidel.
Churchman, C. W. (1979) *The Systems Approach and its Enemies*, New York: Basic Books.
Dubos, R. (1959) *Mirage of Health. Utopias, Progress and Biological Change*, London: George Allen & Unwin.
Ingrosso, M. (ed.) (1987) *Dalla prevenzione della malattia alla promozione della salute*, Milan: Franco Angeli.
Jantsch, E. (1980) *The Self-Organising Universe: Scientific and Human Implications of the Emerging Paradigm of Evolution*, New York: Pergamon Press.
Keeney, B. (1983) *Aesthetics of Change*, New York: Guilford Press.
Kuhn, A. (1975) *Unified Social Science. A System-Based Introduction*, Illinois: Dorsey Press.
Laszlo, E. (1973) *Introduction to Systems Philosophy: Toward a New Paradigm of Contemporary Thought*, New York.
Luhmann, N. (1970) *Soziologische Aufklärung: Aufsätze zur Theorie Sozialer Systeme*, Köln-Opladen, Westdeutscher Verlag.
Luhmann, N. and Habermas, J. (1971) *Theorie der Gesellschaft oder Sozialtechnologie: Was leistet die Systemforschung*, Frankfurt am Main: Suhrkamp.
Maturana, H. R. and Varela, F. J. (1988) *The Tree of Knowledge. The Biological Roots of Human Understanding*, Boston/London: Shambhala.
Opp, K. D. (1970) *Kybernetik und Soziologie*, Berlin.
Ornstein, R. and Sobel, D. (1988) *The Healing Brain. A Radical New Approach to Healthcare*, London: Macmillan.
Parsons, T. (1951), *The Social System*, New York: Free Press of Glencoe.
—— (1970) *Social Structure and Personality*, London.
—— (1977) *Social Systems and the Evolution of Action Theory*, New York: Free Press of Glencoe.
—— (1978) *Action Theory and the Human Condition*, New York: Free Press of Glencoe.
—— (1979/1980) 'On theory and metatheory', *Humboldt Journal of Social Relations* 7(1).
Pike, K. L. (1971) *Language in Relation to a Unified Theory of the Structure of Human Behaviour*, The Hague: Mouton.
Rapoport, A. (1968) 'General systems theory', in *International Encyclopedia of the Social Sciences*, New York: Macmillan and Free Press, pp. 452–8.
Sobel, S. (1979) *Ways of Health. Holistic Approaches to Ancient and Contemporary Medicine*, New York and London: Harcourt Brace Jovanovitch.
Turner, V. (1977) 'Process, system and symbol: a new anthropological synthesis', *Daed.* 106(3): 61–80.
Vries, M. de (1985) *Het Behoud van Leven* (The Preservation of Life), Utrecht: Bohn, Scheltema & Holkema.

Wiener, N. (1957) *Cybernetics: Or Control and Communication in the Animal and the Machine*, New York.

2.6 Humanistic–anthropological paradigm

Antonovski, A. (1987) *Unraveling the Mystery of Health; How People Manage Stress and Stay Well*, San Francisco/London: Jossey-Bass Publishers.
Cousins, N. (1979) *Anatomy of an Illness as Perceived by the Patient*, New York: W.W. Norton.
Lusseyran, J. (1965), *And There Was Light*, Edinburgh: Floris.

2.7 Culturological paradigm

Armstrong, D. (1983) *Political Anatomy of the Body. Medical Knowledge in Britain in the Twentieth Century*, Cambridge: Cambridge University Press.
Beck, U. (date n.a.) *Risk Society. Towards a New Modernity*, London: Sage Publications.
Berg, J. H. van den (1970) *Things (For Metabletic Reflections)*, Pittsburgh: Duquesne University Press.
—— (1972) *A Different Existence*, Pittsburgh: Duquesne University Press.
Bolen, J. S. (1985) *Goddesses in Every Woman. New Psychology of Women*, New York: HarperCollins.
—— (1990) *God in Every Man. A New Psychology of Man's Life and Loves*, New York: HarperCollins.
Bourdieu, P. (1977) *Outline of a Theory of Practice*, Cambridge/London/New York/Melbourne: Cambridge University Press.
Campbell, J. (1949) *A Hero with a Thousand Faces*, New York: Bollingen Foundation.
—— (1988) *The Power of the Myth*, New York: Doubleday.
Condren, M. (1989) *The Serpent and the Goddess. Woman, Religion and Power in Celtic Ireland*, San Francisco: Harper & Row.
Crawford, R. (1977) 'You are dangerous to your health: the ideology and politics of victim blaming', *International Journal of Health Services* 7(4): 663–79.
—— (1980) 'Healthism and the medicalisation of everyday life', *International Journal of Health Services* 10(3): 365–88.
Duden, B. (date n.a.) *Woman Beneath the Skin. A Doctor's Patients in 18th Century Germany*, Harvard, MA: Harvard University Press.
Duerr, P. (date n.a.) *Dreamtime: From Wilderness to Civilization*, Oxford: Basil Blackwell.
—— (1990) *Nacktheit und Scham*, Frankfurt am Main: Suhrkamp.
Elias, N. (1969) *Ueber den Prozess der Zivilisation. Soziogenetische und psychogenetische Untersuchungen*, 2 vols, Bern and München: Francke Verlag.
—— (1983) *Power and Civility. The Civilizing Process*, London: Pantheon Books.
—— (1991) *The Society of Individuals*, Oxford: Basil Blackwell.
Foucault, M. (1971) *Madness and Civilization. A History of Insanity in the Age of Reason*, London: Routledge.
—— (1973) *The Birth of the Clinic: An Archeology of Medical Perception*, New York: Vintage Books.
—— (1976) *Mental Illness and Psychology*, New York: HarperCollins.
—— (1990) *Politics, Philosophy, Culture; Interviews and Other Writings*, London: Routledge.
Freidson, E. (1970) *Professional Dominance: The Social Structure of Medical Care*, New York: Atherton Press.

Gleichmann, P., Goudsblom, J. and Korte, H. (1977) *Macht und Zivilization. Materialien zu Norbert Elias' Zivilizationstheorie*, part I, Frankfurt am Main: Surhrkamp

—— (1984) *Macht und Zivilization. Materialien zu Norbert Elias' Zivilizationstheorie*, part II, Frankfurt am Main: Surhrkamp.

Goudsblom, J. (1977) 'Civilisatie, besmettingsangst en hygiëne. Beschouwingen over een aspekt van het Europese civilisatieproces', *Amsterdams Sociologisch Tijdschrift*, pp. 271–300.

Helman, C. (1984) *Culture, Health and Illness. An Introduction for Health Professionals*, Bristol/London/Boston: Wright.

Hurrelmann, K. (1989) *Human Development and Health*, Munich and London: Springer-Verlag.

Illich, I. (1976) *Medical nemesis: the expropriation of health*, New York: Marion Boyars.

Kleinman, A. (1980) *Patients and Healers in the Context of Culture*, Berkeley: University of California Press.

McNeill, H. (1976) *Plagues and Peoples*, New York: Doubleday.

McNeill, W. H. (1980) *The Human Condition: An Ecological and Historical View*, Princeton, NJ.

Mechanic, D. (1978) *Medical Sociology*, New York: Free Press of Glencoe.

Mitscherlich, A. (1966 and 1967) *Krankheit als Konflikt*, vols 1 and 2, Frankfurt am Main: Surhrkamp.

—— (1967) Der Kranke in der modernen Gesellschaft, Köln: Kiperheuer & Witch.

Sontag, S. (1977) *Illness as a Metaphor*, New York: Schocken.

Swaan, A. de (1988) *In Care of the State. Health Care, Education and Welfare in Europe and the USA*, Oxford: Oxford University Press.

—— (1990) *The Management of Normality. Critical Essays in Health and Welfare*, London: Chapman & Hall.

Szasz, Th. (1970a) *Ideology and Insanity*, New York: Doubleday.

—— (1970b) *The Manufacture of Madness*, New York: Harper & Row.

—— (1981) *Sex: Facts, Frauds and Follies*, Oxford: Blackwell.

Turner, J. (1991) *The Structure of Sociological Theory*, Belmont, CA: Wadsworth.

Verburgh, H. S. (1978) *Paradigma's en begripsontwikkeling in de ziekteleer*, Haarlem: De Toorts.

—— (1983) *Nieuw besef van ziekte en ziek zijn. Over veranderingen in het mensbeeld van de medische wetenschap*, Haarlem: De Toorts.

Wenzel, E. (ed.) (1986) *Die Ökologie des Körpers*, Frankfurt am Main: Surhrkamp.

Wilber, K. (1981) *Up from Eden. A Transpersonal View of Human Evolution*, London: Routledge & Kegan Paul.

—— (1983) *Eye to Eye*, New York: Anchor Books.

Zola, I. K. (1983) *Socio-Medical Inquiries. Recollections, Reflections and Reconsiderations*, Philadelphia, PA: Temple University Press.

5   HEALTHY LIFE-STYLE PROGRAMMES

5.1 Life-style and health

Aakster, C. W. (1978) *Socio-Cultural Variables in the Etiology of Health Disturbances – A Sociological Approach*, Groningen: Tjeenk Willink.

Badura, B. (ed.), *Soziale Unterstützung und chronische Krankheit. Zum Stand sozialepidemiologischer Forschung*, Frankfurt am Main: Surhrkamp.

BIBLIOGRAPHY

Bergler, R. (1974) *Sauberkeit, Norm – Verhalten – Persönlichkeit*, Bern/Stuttgart/ Wien: Hans Huber.

Berkman, L. F. and Breslow, L. (1983) *Health and Ways of Living: The Alameda County Study*, New York: Oxford University Press.

Cockerham, W. C., Lüschen, G., Abel, Th. and Kunz, G. (1989) 'Sport, Gesundheitsstatus und Gesundheitskultur', *Medizin Soziologie* 3/4 jgr., 2(1989) + 1(1990), pp. 62–80.

Dixhoorn, J. van (1990) 'Relaxation therapy in cardiac rehabilitation. A randomised controlled clinical trial of breathing awareness as a relaxation method in the rehabilitation after myocardial infarction', diss., Rotterdam: Erasmusuniversiteit.

Gould, K. L. *et al.* (1992) 'Improved stenosis geometry by quantitative coronary arteriography after vigorous risk factor modification', *American Journal of Cardiology* 69(10): 845–53.

House, J. S. (1980a) 'Zum sozialepidemiologische Verstandnis von Public Health: soziale Unterstützung und Gesundheit', in B. Badura *et al.* (eds) *Zukunftsaufgabe Gesundheits-förderung*, Landesverband der Betriebskrankenkassen in Berlin, Berlin, pp. 173–84.

—— (1980b) *Occupational Stress and the Mental and Physical Health of Factory Workers*, USA: Institute of Social Research.

Lafaille, R. (1984) 'Self-help as self-care', in S. Hatch and I. Kickbusch (eds), *Self-Help and Health*, Copenhagen: WHO, pp.169–76.

Lafaille, R., Geelen, K., van Aalderen, H., Cuvelier, F., van Dijk, P., Janssens, H., Lambooij-Clabbers, E. and Severne, L. (eds) (1981–5) *Zelfhulptechnieken* (Self-Care Techniques), loose-leaf edition, 3 vols, Deventer/Antwerp: Van Loghum Slaterus.

McGann, B. (1980) *Behaviour, Health and Lifestyle*, Dublin: Villa Books.

Niehoff, J.-U. and Wolters, P. (1990) *Ernährung und Prävention. Köpergewichte – ein Beispiel präventionstheoretischer Probleme*, Berlin: Wissenschaftszentrum Berlin für Sozialforschung.

Ornish, D. (1990) *Program for Reversing Heart Disease*, New York: Ballantine Books.

Ornish, D., *et al.* (1990) 'Can lifesyle reverse coronary heart disease?', *The Lancet* 336: 129–33.

Pelletier, K. R. (1977) *Mind as Slayer. A Holistic Approach to Preventing Stress Disorders*, Boston/Sydney: George Allen & Unwin.

—— (1979) *Holistic Medicine. From Stress to Optimum Health*, New York: Delta / Seymour Lawrence.

—— (1981) *Longevity. Fulfilling Our Biological Potential*, New York: Delta / Seymour Lawrence.

—— (1984) *Healthy People in Unhealthy Places. Stress and Fitness at Work*, New York: Delacorte Press / Seymour Lawrence.

Pitts, M. and Phillips, K. (1991) *The Psychology of Health. An Introduction*, London: Routledge.

Setting Priorities (date n.a.), Leiden: NIPG Publications.

Sobel, M. E. (1982) *Lifestyle and Social Structure. Concepts, Definitions, Analyses*, New York: Academic Press.

Veenhoven, R. (1984) *Conditions of Happiness*, Dordrecht: Reidel.

—— (ed.) (1989) *How Harmful is Happiness?*, Rotterdam: Universitaire Pers Rotterdam.

5.2 History of healthy life-style programmes

Flanagan, S. (1989) *Hildegard of Bingen, 1098–1179. A Visionary Life*, London/New York: Routledge.

Gandhi, M. K. (1923) *A Guide to Health*, Madras: Ganesan.

Green, R. M. (1951) *A Translation of Galen's Hygiene (De Sanitate Tuenda)*, Springfield, Illinois: Charles C. Thomas.

Jones, W. H. S. (1947) *Hippokrates. With an English translation*, London: Heinemann / Cambridge, MA: Harvard University Press.

Kristeller, P. O. (1945) 'The school of Salerno, its development and its contribution to the history of learning', in *Bulletin of the History of Medicine* 17: 138–94.

—— (1967) 'Sources of Salernitan medicine in the 12th century', in *Salerno, Civitas Hippocratica*, Jan.–Feb., vol. 1, nos. I–II, pp. 19–26.

Lafaille, R. and Hiemstra, H. (1990) 'The Regimen of Salerno, a contemporary analysis', *Health Promotion International* 5(1): 57–74.

Leslie, C. (1976) *Asian Medical Systems: A Comparative Study*, Berkeley, CA: University of California Press.

Loux, F. R. (1979) *Le Corps dans la société traditionelle*, Paris: Berger-Levault.

Renzi, S. de *et al.* (1852–9) *Collectio Salernitana*, Bologna: Forni Editore.

Rosen, G. (1974) From Medical Police to Social Medicine: Essays on the History of Health Care, New York: Science History Publications.

Schipperges, H. (1985) *Der Garten der Gesundheit. Medizin im Mittelatler*, München: Artemis-Verlag.

—— (1990a) *Heilkunst als Lebenskunde oder die Kunst, vernünftig zu leben. Zur Theorie der Lebensordnung und Praxis der Lebensführung*, Stuttgart: Gesellschaft für Gesundheitsbildung.

—— (1990b) *Hildegard of Bingen*, USA: Philos. Lit.

Siegel, R. E. (1968) *Galen's System of Physiology and Medicine. An Analysis of his Doctrines and Observations on Bloodflow, Respiration, Humors and Internal Diseases*, Basel/New York: S. Karger.

Willett Cummins, P. (1976) *A Critical Edition of Le Régime très utile et très proufitable pour conserver et garder la santé du corps humain*, Chapel Hill: North Carolina Studies in the Romance Languages and Literatures.

5.3 Non-Western healthy life-style programmes

5.3.1 Yoga and Ayurveda

Brosse, Th. (1963) *Etudes Instrumentales des techniques du yoga*, Paris: Ecole française d'Extrême Orient.

Ebert, D. (1986) *Physiologische Aspekte der Yoga*, Leipzig: VEB Georg Thieme.

Funderburk, J. (1977) *Science Studies Yoga. A Review of Physiological Data*, Himalayan International Institute of Yoga Science & Philosophy of USA.

Jaggi, O. O. P. (1979) *Yogi and Tantric Medicine*, Delhi/Lucknow: Atma Ram & Sons.

Medical and Psychological Scientific Research on Yoga and Meditation. An Introduction and Complete List of 452 Scientific Reports (1978) Copenhagen: Peo, Scandinavian Yoga and Meditation School.

Monro, R., Ghosh, A. K. and Kalish, D. (1989) *Yoga Research Bibliography. Scientific Studies on Yoga and Meditation*, London: Yoga Biomedical Trust.

5.3.2 Meditation

Leuner, H. C. (1984) *Guided Affective Imagery. Mental Imagery in Short Term Psychotherapy*, USA: Thieme Medical Publications.

Murphy, M. and Donovan, S. (1988) *The Physical and Psychological Effects of Meditation: A Review of Contemporary Meditation Research with a Comprehensive Bibliography 1931–1988*, San Rafael: Esalen Institute.

Shapiro, D. H. and Walsh, R. W. (eds) (1984) *Meditation: Contemporary and Classical Perspectives*, New York: Aldine.

Staal, F. (1975) *Exploring Mysticism*, Harmondsworth, Middlesex: Penguin.

5.4 Health promotion (WHO)

Anderson, R. (1984) 'Health promotion: an overview', *European Monographs in Health Education Research* 6: 1–76.

Dean, K. (1983) *Influence of Health Beliefs on Lifestyles: What Do we Know?*, Copenhagen: WHO Regional Office for Europe.

Health Promotion. An International Journal, Oxford: Oxford University Press.

Health Promotion. Concepts and Principles in Action. A Policy Framework (undated), Copenhagen: WHO, Regional Office for Europe.

Health Promotion: A Discussion Document on the Concept and Principles (1984), Copenhagen: WHO Regional Office for Europe.

Health Promotion: Concepts and Principles (1984), Copenhagen: WHO Regional Office for Europe.

Health Promotion in the Working Field (1987), Cologne: Federal Centre for Health Education / Copenhagen: WHO Regional Office for Europe.

Intervention Studies Related to Lifestyles Conductive to Health (1983), Copenhagen: WHO Regional Office for Europe.

Kaplun, A. and Wenzel, E. (eds), *Health Promotion in the Working World*, Berlin: Springer.

Kickbusch, I. (1981) 'Involvement in health: a social concept', *International Journal of Health Education* 24(4): 3–15.

—— (1986) 'Health promotion: a global perspective', *Canadian Journal of Public Health* 77: 321–6.

—— (1989) *Good Planets are Hard to Find*, WHO Healthy Cities Papers No. 5, Copenhagen: Fadl.

Liederkerken, P. C. *et al.* (eds) (1990) *Effectiveness of Health Education*, Utrecht: Dutch Health Education Centre.

Lifestyles and their Impact on Health (1982) Copenhagen: WHO Regional Office for Europe, Copenhagen.

Mahler, H. (1981) 'The Meaning of "Health for All by the Year 2000" ', *World Health Forum* 2(1): 5–22.

Nutbeam, D. (1985) *Health Promotion Glossary*, Copenhagen: WHO Regional Office.

O'Neill, P. (1983) *The Meaning of 'Health for All by the Year 2000'*, London: WHO.

Positive Health, Journal of the Institute for Health Promotion, Cardiff, Wales: University of Wales, College of Medicine.

Regional Programme in Health Education and Lifestyles (1981) Copenhagen: WHO Regional Office for Europe.

Robertson, J. (date n.a.) *Lifestyle and Health: Different Possible Scenarios*, Copenhagen: WHO Regional Office for Europe.

Roscam Abbing, H. (1979) *International Organizations in Europe and the Right to Health Care*, Deventer: Kluwer.

'The Adelaide recommendations on healthy public policy' (1988), *Health Promotion: an International Journal* 3(2): 183–6.

'The Ottawa Charter for Health Promotion' (1986), *Health Promotion: an International Journal* 1(4): III–V.

Wenzel, E. (1983) 'Health promotion and lifestyles', *Hygiene* 1: 57–61.

WHO (1958) *The First Ten Years of the World Health Organization*, Geneva: WHO.

—— (1981) *Global Strategy for Health for All by the Year 2000*, Geneva: WHO (Health for All Series, No. 3).

6 COMMUNITY AND HEALTH

Badura, B. (1981) *Soziale Unterstützung und chronische Krankheid*, Frankfurt am Main: Sürhrkamp.

Brown, R. E. (1991) 'Community action for health promotion: a strategy to empower individuals and communities', *International Journal of Health Services* 21 (3): 441–56.

Caplan, G. (1974) *Support Systems and Community Mental Health*, New York: Academic Press.

Cobb, S. (1976) 'Social support as a moderator of life stress', *Psychosomatic Medicine*, 38: 300–14.

Cochran, M. (1987) 'Empowering families: an alternative to the deficit model', in K. Hurrelmann, F. X. Kaufmann and F. Lösel (eds) *Social Intervention: Potential and Constraints*, Berlin: De Gruyter, 105–20.

Cohen, S. and Syme, S. L. (eds) (1985) *Social Support and Health*, Orlando: Academic Press.

Froland, C., Pancoas, D. L., Chapman, N. K. and Kimboko, P. (1981) *Helping Networks and Human Services*, Beverly Hills: Sage.

Gerhards, U. (1982) 'Coping and social action', *Sociology of Health and Illness* 1: 195–225.

Gottlieb, B. H. (1981) *Social Networks and Social Support*, Beverly Hills and London: Sage.

Heller, K. and Swindle, R. W. (1983) 'Social networks, perceived social support, and coping with stress', in R. D. Felner *et al.* (eds), *Preventive Psychology*, New York: Pergamon, 87–103.

House, J. S. (1981) *Work Stress and Social Support*, Reading: Addison-Wesley.

Levin, L. S. (1983) 'Lifestyle research and health promotion policy with special reference to mediating structures', in Scottish Health Education Group (eds), *European Monograph in Health Education Research*, pp. 19–26.

7 ENVIRONMENT AND HEALTH

Lovelock, J. (1979) *A New Look at Life on Earth*, Oxford: Oxford University Press.

Merchant, C. (1988) *The Death of Nature: Women, Ecology and the Scientific Revolution*, New York: Harper & Row.

Whisehunt, D. W. (1974) *The Environment and the American Experience: A Historian Looks at the Ecological Crisis*, Port Washington, NY: Kennikat Press.

WHO (1989) World Health Assembly Resolution WHA 42.26 (19 May 1989), *WHO's Contribution to the International Efforts Towards Sustainable Development*.

Wigley, T. M. L., Ingram, M. J. and Farmer, G. (eds) (1981) *Climate and History: Studies in Past Climates and Their Impact on Man*, Cambridge: Cambridge University Press.

World Commission on Environment and Development (1987) *Our Common Future*, Oxford: Oxford University Press.

8 HEALTH POLICY

Galbraith, K. (1992) *The Culture of Contentment*, USA.

McKinlay, J. B. (1984) *Issues in the Political Economy of Medical Care*, New York: Methuen / London: Tavistock Publications.

Milio, N. (1986) *Promoting Health Through Public Policy*, Ottawa: Canadian Public Health Association.

Salmon, J. W. (1990) *The Corporate Transformation of Health Care*, Part I: *Issues and Directions*, Amityville, NY: Baywoord.

—— (1992) *The Corporate Transformation of Health Care*, Part II: *Reflections and Implications*, Amityville, NY: Baywoord.

Salmon, J. W. and Göpel, E. (1990) *Community Participation and Empowerment Strategies in Health Promotion*, 7 vols, Bielefeld: Zentrum für Interdisziplinare Forschung.

9 METHODOLOGY

9.1 Qualitative research

9.1.1 Biographical methodology

Allport, G. W. (1942) *The Use of Personal Documents in Psychological Science*, New York: Social Science Research Council.

Arbeitsgruppe Bielefelder Soziologen (ed.) (1973) *Alltagswissen, Interaktion und gesellschaftliche Wirklichkeit*, Bd. 1: *Symbolischer Interaktionismus und Ethno-methodologie*, Bd. 2 *Ethnotheorie und Ethnografie des Sprechens*, Reinbek bei Hamburg.

—— (ed.) (1976) *Kommunikatieve Sozialforschung*, München.

Bios. Zeitschrift für Biographieforschung und Oral History, Fernuniversität Hagen, Postfach 940, 5800, Hagen: Leske & Budrick.

Cook, T. D. and Reichardt, CS. (1979) *Qualitative and Quantitative Methods in Evaluation Research*, Beverly Hills: Sage.

Cremer-Schäfer, H. (1985) *Biographie und Interaktion*, München: Profil-Verlag.

Denzin, N. K. (1978) *The Research Act in Sociology: A Theoretical Introduction to Sociological Methods*, London: Butterworth.

Fuchs, W. (1984) *Biographische Forschung: Eine Einführung in Praxis und Methoden*, Opladen: Westdeutscher Verlag.

Glaser, B. and Strauss, A. (1967) *The Discovery of Grounded Theory: Strategies of Qualitative Research*, Chicago: Aldine.

Hampson, S. E. (1982) *The Construction of Personality. An Introduction*, London: Routledge.

Human Studies. A Journal for Philosophy and the Social Sciences. Norwood, New Jersey: Alex Publishing Corporation.

Jütteman, G. and Thomae, H. (1987) *Biographie und Psychologie*, Berlin: Springer-Verlag.

Kohli, M. (1978) *Soziologie des Lebenslauf*, Darmstadt/Neuwied: Luchterhand.

—— (1981) 'Wie es zur biographischen Methode kam und was daraus geworden ist', *Zeitschrift für Soziologie* 10: 272–93.

—— (1983) 'Biographieforschung im Deutschen Sprachbereich', *ASI-News* 6: 5–32.

Kruse, A. (1987) 'Biographische Methode und Exploration', in G. Jütteman, H. Thomae (eds), *Biographie und Psychologie*, Berlin/Heidelberg/New York/London/Paris/Tokyo: Springer-Verlag.

Lachmund, J. and Stollberg, G. (1987) 'Zur medikalen Kultur des Bildungsbürgertums um 1800. Eine soziologische Analyse anhand von Autobiographien', *Jahrbuch des Instituts für Geschichte der Medizin der Robert Bosch Stiftung*, Band 6, Stuttgart: Hippokrates Verlag, pp. 163–84.

Lafaille, R. and Lebeer, J. (1989) *The Study of Life Histories for Understanding Health and Healing. The Observation of the Biographical Process*. 2 vols, Pre-Publication International Network for a Science of Health, Antwerp: International Institute for Advanced Health Studies.

—— (1991) 'The relevance of life histories for understanding health and healing', *Advances* 7(4): 16–31.

Lafaille, R., Lebeer, J. and Tilkin-Franssens, D. (1991) *Some Current Scientific Models for Biographical Research in the Health Sciences*, Pre-Publication International Network for a Science of Health, Antwerp: International Institute for Advanced Health Studies.

Lamnek, S. (1987) *Qualitative Sozialforschung*, Band 1: *Methodologie*, München and Weinheim: Psychologie Verlags Union.

Miles, M. B. and Huberman, M. A. (1984) *Qualitative Data Analysis. A Source Book of New Methods*, London: Sage Publications.

Plummer, K. (1983) *Documents of Life: An Introduction to the Problems and Literature of a Humanistic Method*, London: Allen & Unwin.

Ricoeur, P. (1965) *History and Truth*, Evanston, ILL: Northwestern University Press.

Spence, D. P. (1982) *Narrative Truth and Historical Truth: Meaning and Interpretation in Psychoanalysis*, New York: W. W. Norton.

Vansina, J. (1972) *Oral History: A Study in Historical Methodology*, London: Routledge & Kegan Paul.

9.1.2 Theories of identity, the self and the life cycle

Brose H.-G. (1988) 'Von Ende des Individuums. Zur Individualität ohne Ende', *Biographie und Gesellschaft*, Band 4, Opladen: Leike & Budrick.

Gergen, K. J. (1991) *The Saturated Self. Dilemmas of Identity in Contemporary Life*, New York: Basic Books.

Giddens, A. (1991), *Modernity and Self-Identity: Self and Society in the Late Modern Age*, Cambridge: Polity Press.

Hampson, S. E. (1988) *The Construction of Personality*, London: Routledge.

Hardy, J. (1987) *A Psychology with a Soul: Psychosynthesis in Evolutionary Context*, London: Routledge & Kegan Paul.

Honess, T. and Yardley, K.M. (1987) *Self and Identity. Perspectives Across the Lifespan*, London: Routledge.

9.2 New models of quantitative research

Arnodl, Vl. (1983) *Catastrophe Theory*, Berlin/Heidelberg: Springer-Verlag.

Babloyantz, A. (1988) 'Chaotic dynamics in brain activity', in E. Bazar (ed.) *Dynamics of Sensory and Cognitive Processes by the Brain*, Berlin: Springer-Verlag, 196–202.

Canter, D. (1985) *Facet Theory: Approaches to Social Research*, Heidelberg: Springer-Verlag.

Cvitanovic, P. (1988) *Universality in Chaos*, Bristol: Hilger.

Glass, L. and Mackey, M. (1988) *From Clocks to Chaos – The Rhythms of Life*, New York: Princeton University Press.

Gleick, J. (1988) *Chaos*, London: Heinemann.

Goldbergher, A. (1987) 'Non-linear dynamics, fractals, cardiac physiology and sudden death', in L. Rensing and Van Der Heiden (eds), *Temporal Disorder in Human Oscillatory Systems*, Berlin: Springer-Verlag.

Janssens, L. (1990) *The Integral-Differential Equations on the Ventricular Cardiac Suction Pump. A Catastrophe Theoretic Approach to the Matching Stimulation Method*, diss., Antwerp: University of Antwerp.

Katz, J. (1972) *Experimentation with Human Beings. The Authority of the Investigator, Subject, Professions, and State in the Human Experimentation process*, New York: Russell Sage Foundation.

Lauwerier, H. A. (1987) *Fractals*, Amsterdam: Aramith Uitgevers.

Mandelbrot, B. (1982) *The Fractal Geometry of Nature*, New York: W. H. Freeman.

Oud, J. H. L. (1978) *Systeem-methodologie in sociaal-wetenschappelijk onderzoek*, Nijmegen: Alfa.

Peitgen, H.-O. and Richter, P. H. (1986) *The Beauty of Fractals*, Berlin: Springer-Verlag.

Poston, T. and Stewart, I. (1977) *Catastrophe Theory and its Applications*, London: Pitman.

Rapp, P. E., Bachore, T. R. and Zimmerman, I. D. (1990) 'Dynamical characterisation of brain electrical activity', in S. Krasner (ed.) *Ubiquity of Chaos*, New York: AAAS.

Reason, P. and Rowan, J. (eds) (1981) *Human Inquiry: A Sourcebook of New Paradigm Research*, Chichester: Wiley.

Saunders, P. T. (1980) *Introduction to Catastrophe Theory*, Cambridge: Cambridge University Press.

Schaffer, W. M. and Kot, M. (1985) 'Nearly one dimensional dynamics in an epidemic', *Journal of Theoretical Biology* 112: 408–27.

Schroeder, M. (1991) *Fractals, Chaos, Power Laws: Minutes From an Infinite Paradise*, New York: Freeman.

Thom, F. (1972) *Stabilité structurelle et morphogénèse*, New York: Benjamin.

Zeeman, E. C. (1977) *Catastrophe Theory. Selected Papers 1972–1977*, London: Addison-Wesley.

10 RELATED THEMES

10.1 Preventive medicine, self-help and self-care

Forbes, A. (1976) *Try Being Healthy*, Holsworthy: Health Science Press.

Gann, R. (1986) *The Health Information Handbook*, London: Gower.

Inglis, B. and West, R. (1983) *The Alternative Health Guide*, London: Michael Joseph.

Null, G. (1984) *The Complete Guide to Health and Nutrition*, London: Arlington.

10.2 Psychosomatics and mind–body relationships

Advances. The Journal of Mind–Body Health, Michigan: Fetzer Institute.

Antonovsky, A. (1980) *Health, Stress and Coping*, San Francisco: Jossey-Bass.

—— (1987) *Unravelling the Mystery of Health. How People Manage Stress and Stay Well*, San Francisco: Jossey Bass.

Brennan, B. A. (1988) *Hands of Light. Guide to Healing Through the Human Energy Field. A New Paradigm for the Human Being in Health, Relationship, and Disease*, Toronto/New York/London/Sydney/Auckland: Bantam Books.

Dam, F. van (1989) 'Does happiness heal? The case of fighting cancer with hope', in R. Veenhoven (ed.), *How Harmful is Happiness?*, Rotterdam: Universitaire Pers, Rotterdam.

Fisher S. (1986) *Development and Structure of the Body Image*, 2 vols, London: Lawrence Erlbaum Associates.

Gaarder, K. R. (1977) *Clinical Biofeedback: A Procedural Manual*, Baltimore, MD: The Williams & Wilcins Co.

Groddeck, G. (date n.a.) *Meaning of Illness. Selected Psychoanalytical Writings Including Correspondence with Sigmund Freud*, London: H. Karnac Publishers.

Grof, S. (1985) *Beyond the Brain. Birth, Death and Transcendence in Psychotherapy*, Albany: State University of New York Press.

Grossarth-Maticek, R., Bastiaans, J. and Kanazir, D. T. (1985) 'Psychosocial factors as strong predictors of mortality from cancer, ischemic heart disease and stroke: the Yugoslav prospective study', *Journal of Psychosomatic Research* 29: 167–76.

Kaplun, A. (ed.) (1989) *Health Promotion and Chronic Illness. Discovering a New Quality of Health*, Köln: Federal Centre for Health Education.

Lazarus, R. S. and Folkman, S. (1984) *Stress, Appraisal, and Coping*, New York: Springer.

Lerner, M. (1985) 'A report on complementary cancer therapies', *Advances. The Journal of Mind–Body Health* 2: 31–43.

LeShan, L. (1977) *You Can Fight for Your Life: Emotional Factors in the Treatment of Cancer*, New York: M. Evans & Co.

—— (1982) *The Mechanic and the Gardener. Making the Most Out of the Holistic Revolution in Medicine*, New York: Holt/Reinhart/Wiston.

—— (1989) *Cancer as a Turning Point. A Handbook for People with Cancer, their Families, and Health Professionals*, New York: E. P. Dutton.

—— (1990) *The Dilemma of Psychology*, USA: a Button Book.

Lynch, J. J. (1977), *The Broken Heart: The Medical Consequences of Loneliness*, New York: Basic Books.

—— (1979) *The Language of the Heart*, New York: Basic Books.

—— (1982) 'Interpersonal aspects of blood pressure', *Journal of Nervous and Mental Diseases* 170: 143–53.

Milz, H. (1992) *Der Wiederentdeckte Körper. Vom schöpferischen Umgang mit sich Selbst*, München–Zürich: Artemis & Winkler.

Morris, D. B. (1991) *The Culture of Pain*, Berkely: University of California Press.

Prince, R. (ed.) (1982) 'Shamans and endorphin', special issue of *Ethos*, Winter, 10(4).

Remen, N. (1980) *The Human Patient*, New York: Anchor Press/Doubleday.

Rijke, R. P. C. (1985) 'Cancer and the development of will', *Theoretical Medicine* 6: 133–42.

Seleye, H. (1974) *Stress Without Distress*, Philadelphia/NewYork: J. P. Lippincott.
—— (1978) *The Stress of Life*, New York: McGraw-Hill.
Simonton, O., Matthews-Simonton, S. and Creighton, J. (1981) *Getting Well Again*, Toronto/New York/London: Bantam Books.
Vrancken, M. (1989) *Chronische Pijn, Het kruis van de geneeskunde* (Chronic Pain: the Cross of Medicine), diss., Rotterdam: Erasmusuniversiteit.
Weizsäcker, V. von (date n.a.) *Unity of Nature*, New York: Farrar Straus and Giroux.

10.4 Placebo research

Jospe, M. (1980) *The Placebo Effect in Healing*, Toronto: Toronto-Lexington Books.
Martens, F. (1984) 'Effet placebo et transfert', *Psycho-analyse* 1(1): 32–62.
'Placebo effects', Special Issue of *Advances*, 1986, 1(2).
Shapiro, A. K. (1960) 'A contribution to the history of the placebo effect', *Behav. Sci.* 5: 109.
—— (1963) 'Psychological use of medication', in M. Lief (ed.), *Psychological Basis of Medical Practice*, New York: Harper & Row.
—— (1971) 'Placebo effects in medicine, psychotherapy and psychoanalysis', in A.E. Bergin and S. L. Garfield (eds), *Handbook of Psychotherapy and Behavioral Change*, New York: Wiley, pp. 439–73.

10.5 Complementary and traditional medicine

Argument-Sonderband (eds) (1983) *Alternative Medizin*, Argument-Sonderband AS 77, Berlin: Argument Verlag.
Bannerman, R. H., Burton, J. and Wen-Chieh, C. (1983) *Traditional Medicine and Health Care Coverage*, Geneva: WHO.
Dijk, P. van (1979) *Geneeswijzen in Nederland en Vlaanderen* (Complementary Medicine in Holland and Flanders; in Dutch), Deventer: Ankh-Hermes.
—— (1981) *Volksgeneeskunst in Nederland en Vlaanderen* (Folk Medicine in Holland and Flanders), Deventer: Ankh-Hermes.
Dossey, L. (1982) *Space, Time and Medicine*, London: Routledge & Kegan Paul.
Foster, G. M. (1984) *Medical Anthropology*, New York: Wiley.
Fulder, S. (1989) *The Handbook of Complementary Medicine*, London: Coronet Books.
Hastings, A. C. (ed.) (1981) *Health for the Whole Person*, Boulder, CO: Westview Press.
Helman, C. (1985) *Culture, Health and Illness*, Bristol: Wright.
Jenerick, H. P. (ed.) (1973) *Proceedings of the National Institute of Health Acupuncture Research Conference, Maryland*, Bethesda: Department of Health, Education and Welfare (NIH), Publication, No. 74-165.
Kapchuk, T. (1984) *The Web that Has No Weaver*, London: Hutchinson.
Kenyon, J. (1986) *21st Century Medicine: A Layman's Guide to the Medicine of the Future*, Wellingborough: Thorsons.
Letwith, G. T. (ed.) (1985) *Alternative Therapies. A Guide to Complementary Medicine for the Health Professional*, London: Heinemann.
Needham, J. and Gwie-Djen, L. (1980) *Celestial Lancets: A History and Rationale of Acupuncture and Moxa*, Cambridge: Cambridge University Press.
Pokert, M. (date n.a.)*The Theoretical Foundation of Chinese Medicine: Systems of Correspondence*, Cambridge, MA: MIT Press.

Salmon, J. W. (ed.) (1984) *Alternative Medicines: Popular and Policy Perspectives*, New York: Methuen Inc. / London: Tavistock Publications.

Weil, A. (1983) *Health and Healing: Understanding Conventional and Alternative Medicine*, Los Angeles: Houghton Mifflin.

West, R. and Trevelyan, J. E. (1985) *Alternative Medicine: A Bibliography of Books in English*, London: Mansell.

10.6 Critiques of conventional medicine

Carlson, R. J. (1975) *The End of Medicine*, New York: Wiley-Interscience.

Carter, C. O. and Peel, J. (1969) *Equalities and Inequalities in Health*, New York: Academic Press.

Fulder, S. (1987) *How to Survive Medical Treatment*, London: Century-Hutchinson.

Mellvill, A. and Johnson, C. (1982) *Cured to Death, the Effect of Prescription Drugs*, London: Secker & Warburg.

Weitz, M. (1980) *Health Shock: How to Avoid Ineffective and Hazardous Medical Treatment*, Englewood Cliffs, NJ: Prentice-Hall / London: David and Charles.

NAME INDEX

Abraham, A., 38
Ackoff, R., 170, 177
Adelaide Recommendations, xi, 161
Africa, 144
Alma Ata Declaration, xv, 161
Amro Bank, 153
Anderson, E., 85, 106
Andolfi, M., 86
Andrews, V., 210
Anthroposophical Medical Association, 52
Antonovsky, A., 77, 80, 165, 166, 207, 209, 214
Aral Sea, 144
Argument-Sonderband, 160
Aristotle, 34
Asia, 144
Assagioli, R., 138, 215

Baade, W., 70
Babloyantz, A., 66
Baretta, G., 242
Barlow, K., 51
Batenburg, J., 242
Bateson, G., 6, 7, 85, 86, 94, 95, 98
Beck, U., 12, 28
Becker, H., 9
Berg, van den, J., 12
Bergsma, J., 80
Berliner, H., 160
Bertalanffy, L. von, 5, 6, 85, 129, 133
Bichat, X., 38, 39
Bickel, H., 5
Bierstadt, R.,170
Bismarck, O. von, 206
Blattner, B., 160
Blok, K., 151, 153
Bohm, D., 67, 172

Bohr, N., 67
Bott, V., 230
Boulding, K., 6
Bourdieu, P., 12
Boven, R., 110
Bowen, M., 96
Brazil, 154
Brenner, M., 177
Breugelink, G., 147
Brown, E., 126, 127
Brown, L., 144, 145, 146, 169
Brown, R., 163
Brundtland, G., 149
Burghgraeve, P., 85
Bush, G., 162

Callebaut, D., xvi, xxi, 6, 59
Callebaut, W., 7
Campbell, J., 12
Canadian School, 9
Cannon, W., 139
Canter, D., xix, xx, xxi, 115, 183, 185, 188
Capra, F., xiv, 2, 10, 70, 160, 172
Carlson, R., 160
Cassandra, 124
Centre for Social Development, 246
Cereseto, M., 174
China, 107
Churchill, W., 129
Churchman, C., 170, 177
Clabbers, E., 10
CLTM, 146, 148, 153
Coenen, H., 9, 242
Columbia, 152
Compernolle, T., 8
Copernicus, N., 69
Copius Peereboom, J., 147, 148

SUBJECT INDEX